For Grade **4**

Science on Target

Using Graphic Organizers to Improve Science Skills

This Book Belongs To:

Written By:
Andrea Karch Balas, Ph.D

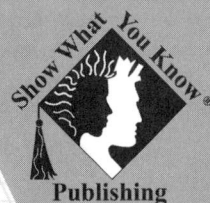

Show What You Know®
Publishing

Published By:

Show What You Know® Publishing

A Division of Englefield & Associates, Inc.

P.O. Box 341348

Columbus, OH 43234-1348

1-877-PASSING (727-7464)

www.showwhatyouknowpublishing.com

Printed in the United States of America

11 10 09   20 19 18 17 16 15 14 13 12 11 10 9 8 7 6 5 4 3 2 1

ISBN: 1-59230-330-7

About the Author

Andrea Karch Balas, Ph.D., is an educator and a scientist who has taught both in the traditional classroom and in nonformal educational settings, from kindergarten to adult. Andrea has presented her research on the teaching and learning of Science both nationally and internationally. Andrea recevied her doctorate in education from The Ohio State University and is currently the General Curriculum Coordinator in a private K–12 school. In addition to this book, Andrea is the co-author of the Ready, Set, Show What You Know® series for grades K–3 in Ohio and Florida.

Acknowledgements

Show What You Know® Publishing acknowledges the following for their efforts in making these skill-building materials available for students, parents, and teachers.

Cindi Englefield, President/Publisher
Eloise Boehm-Sasala, Vice President/Managing Editor
Jill Borish, Production Editor
Christine Filippetti, Production Editor
Trisha Barker, Assistant Editor
Jennifer Harney, Illustrator/Cover Designer

Table of Contents

Visual Glossary of Graphic Organizers

A graphic organizer is an instructional tool used to illustrate prior knowledge about a topic. This visual glossary will give you an idea of what the graphic organizers used in the *Science on Target for Grade 4, Student Workbook* look like, as well as the best way for you to put them to practical use in your everyday Science lessons.

Science on Target

Compare and Contrast Chart

	name 1	name 2
attribute 1		
attribute 2		
attribute 3		

Compare and Contrast Chart uses: Show similarities and differences between two things (people, places, events, ideas, etc.).

Examples include: The comparison of plants and animals. Plants and animals are eukaryotes but plant cells contain additional organelles (cell walls, chloroplasts). Plants make their own food through the process of photosynthesis. Plants are the beginning of the food chain for all animals.

Questions to ask: What are the objects, processes, or procedures being compared? What are the component parts of each? How are they similar? How are they different?

Concept Map

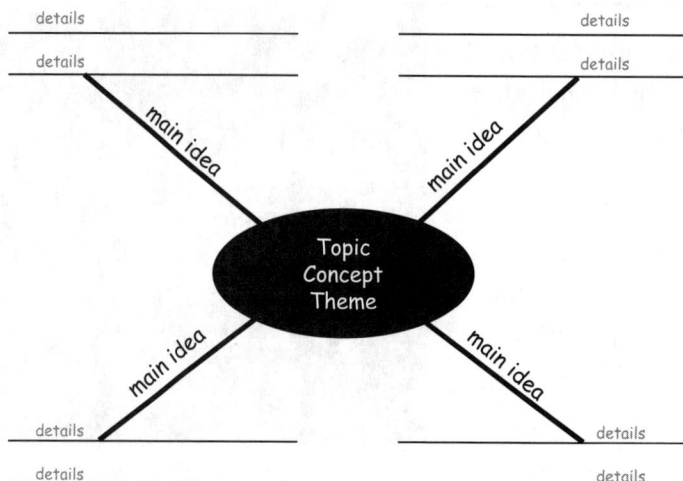

Concept Map uses: Description of a central idea and the relationship of supporting ideas, topics, or functions.

Examples include: If "States of Matter" is the central topic, the connected ideas would be solid state, liquid state, and gaseous state. For each of the states of matter, details could be added to relate the form of the molecules in the state. For example, solids are rigid around a fixed point, and liquids and gases take the shape of their containers.

Questions to ask: How are the ideas connected or interrelated? What details are important to include on the map?

Cycle

Cycle uses: Illustrates the repeated stages or events that occur to create a specific product or event.

Examples include: The illustration of cell reproduction, rock formation, and weather conditions.

Questions to ask: What are the critical stages or events in the cycle? What is the sequence in the activity? Where or what is the event that promotes the continuity of the cycle?

        © **Englefield & Associates, Inc.**

Diagram

Plant Cell

cell wall
cell membrane
mitochondrion

chloroplast
ribosomes
endoplasmic reticulum
nuclear membrane
nucleus
nucleolus
chromosome
vacuole
Golgi apparatus
cytoplasm

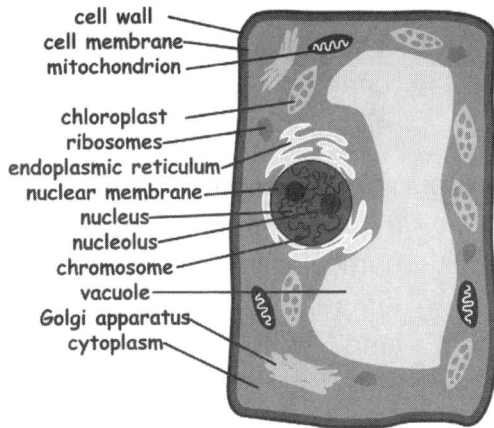

Diagram uses: Provides a tool for describing the relationship between the parts of a system.

Examples include: The model of the cell, the layers of soil, or the components of a circuit.

Questions to ask: What are the parts of the object, theme, or system? What is a visual representation of the information about this process or experiment?

Dichotomous Key

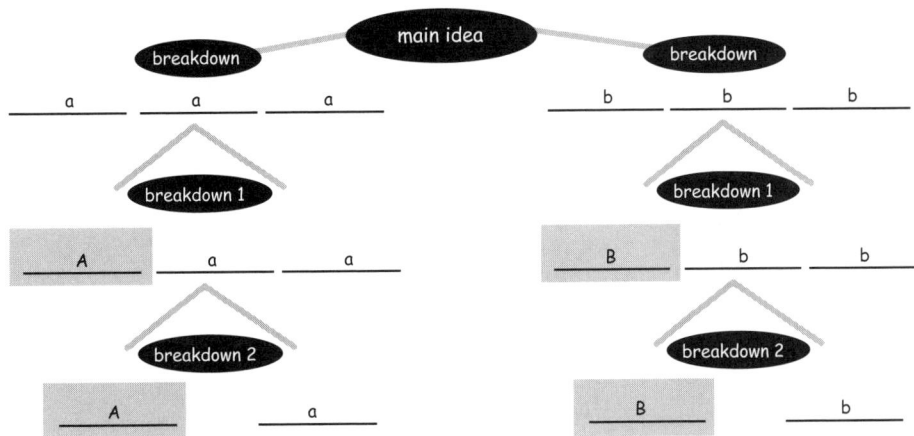

breakdown   main idea   breakdown

a   a   a   b   b   b

breakdown 1   breakdown 1

A   a   a   B   b   b

breakdown 2   breakdown 2

A   a   B   b

Dichotomous Key uses: Characteristics are used to divide a group of objects, organisms, or ideas into two groups until one object remains in each division.

Examples include: Separating organisms into Kingdoms based on attributes.

Questions to ask: How can I separate these objects, organisms, or ideas based on a given trait?

Graph

title

variable b

80
70
60
50
40
30
20
10
0

0   1   2   3   4   5   6

variable a

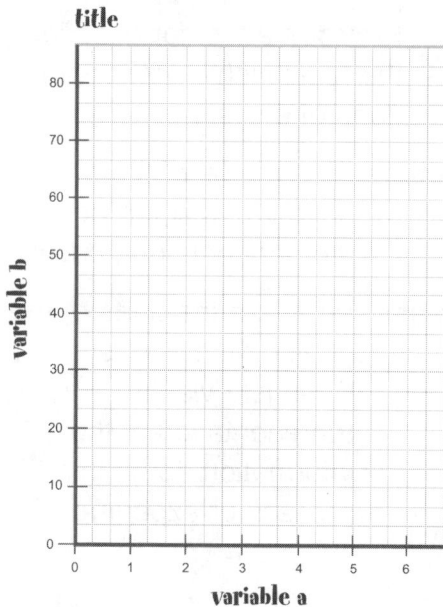

Graph uses: To visually depict collected data.

Examples include: The results of a survey or changes that occur in a given situation.

Questions to ask: Vary depending on the graph, e.g., How many people like a particular fruit? How did the temperature change over time?

Organizational Outline

Title

I. Topic 1
   A. detail
   B. detail

II. Topic 2
   A. detail
   B. detail

Organizational Outline uses: Organizing information, sequencing processes or events.

Examples include: Organizing information from a Science article or presenting information about a mutlifaceted topic such as ecosytems.

Questions to ask: What details are given to support the topic?

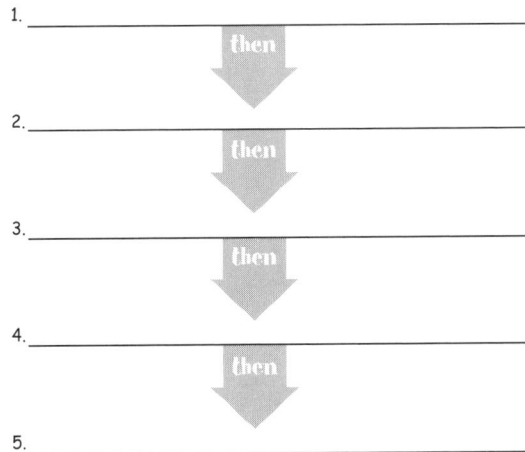

Series of Events Chain or Flow Chart

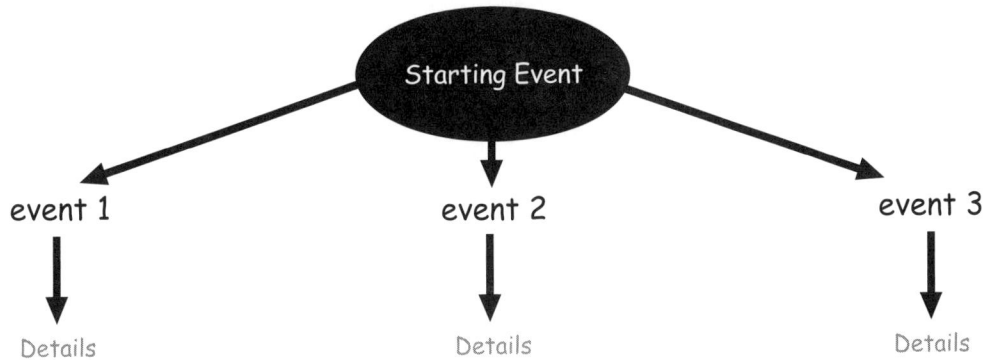

Starting Event

event 1   event 2   event 3

Details   Details   Details

or

Event: _____

1. _____

then

2. _____

then

3. _____

then

4. _____

then

5. _____

**Series of Events Chain
or Flow Chart uses:** Description of stages, steps, sequences, or actions.

Examples include: The formation of something like clouds; the steps in a procedure or experimental design; a sequence of events like erosion or weathering; or the actions leading to the research and development of a product like television or a procedure like space exploration.

Questions to ask: How does the event or process begin? What are the next stages or steps? How are they connected? What is the end product or result?

Timeline

low   high

beginning date   ending date

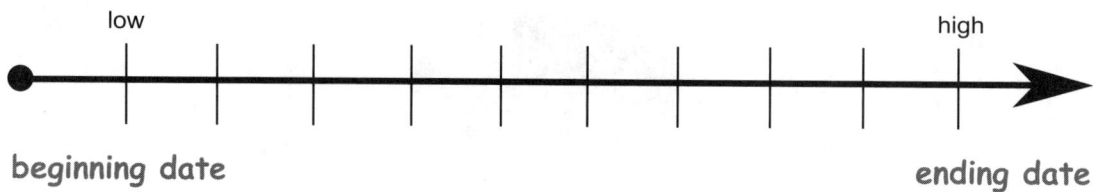

Timeline uses: Showing historical events, creating a record, or charting events.

Examples include: The development of the light bulb, eras of geological time, a growth chart for plants, phases of the moon.

Questions to ask: When does the activity or process begin? When does it end? What happened on a specific date in time?

Venn Diagram

Thing 1   Thing 2

Things that make Thing 1 different from Thing 2

Things that Thing 1 & Thing 2 have in common

Things that make Thing 2 different from Thing 1

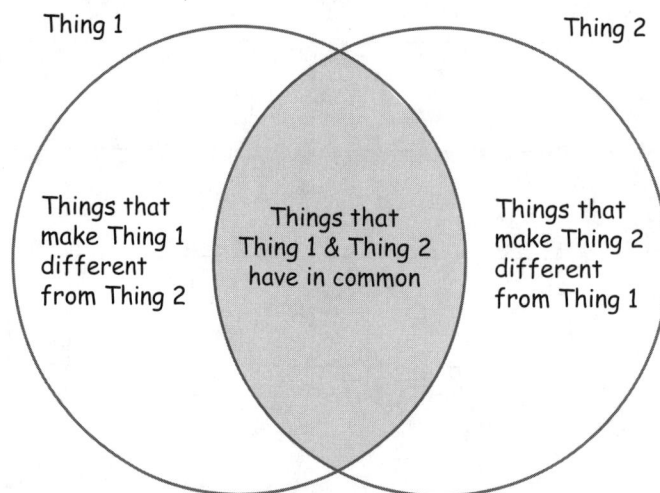

Venn Diagram uses: The comparison and contrast of objects, events, organisms, or themes.

Examples include: The comparison of the microscope and telescope. They both assist the eye in seeing objects. The microscope enlarges smaller objects; the telescope makes objects that are far away closer for observation.

Questions to ask: How are the objects, events, or themes the same? How are they different?

Chapter 1

The activities in this section of the book will focus on Science as Inquiry.

These activities will help you develop the skills necessary to do scientific inquiry and understand scientific inquiry. These activities include:

Science as Inquiry

- asking questions about the world around you,

- planning simple investigations,

- selecting and using scientific tools and equipment,

- using data to explain your observations or experiment results,

- communicating your ideas and results, and

- reviewing the work of other groups.

Use the "Clues for Success" Checklists as you complete each activity in this section as a tool to help you do your best work.

Step 1

Read the scenario "The Marigold Planting Project."

The Marigold Planting Project

The class plants marigold seeds to use as centerpieces for the end-of-the-year party. Each student is given the instructions to fill a yogurt cup with soil and plant 8 seeds in the cup. The soil should be watered with 1/4 cup of water every other day. After two weeks, the results of most of the students were the same but some were quite different.

16 of the students had 8 seedlings in their cup.

7 of the students had 3 seedlings in their cup.

1 student had 15 seedlings in her cup.

5 students had no seedlings in their cup.

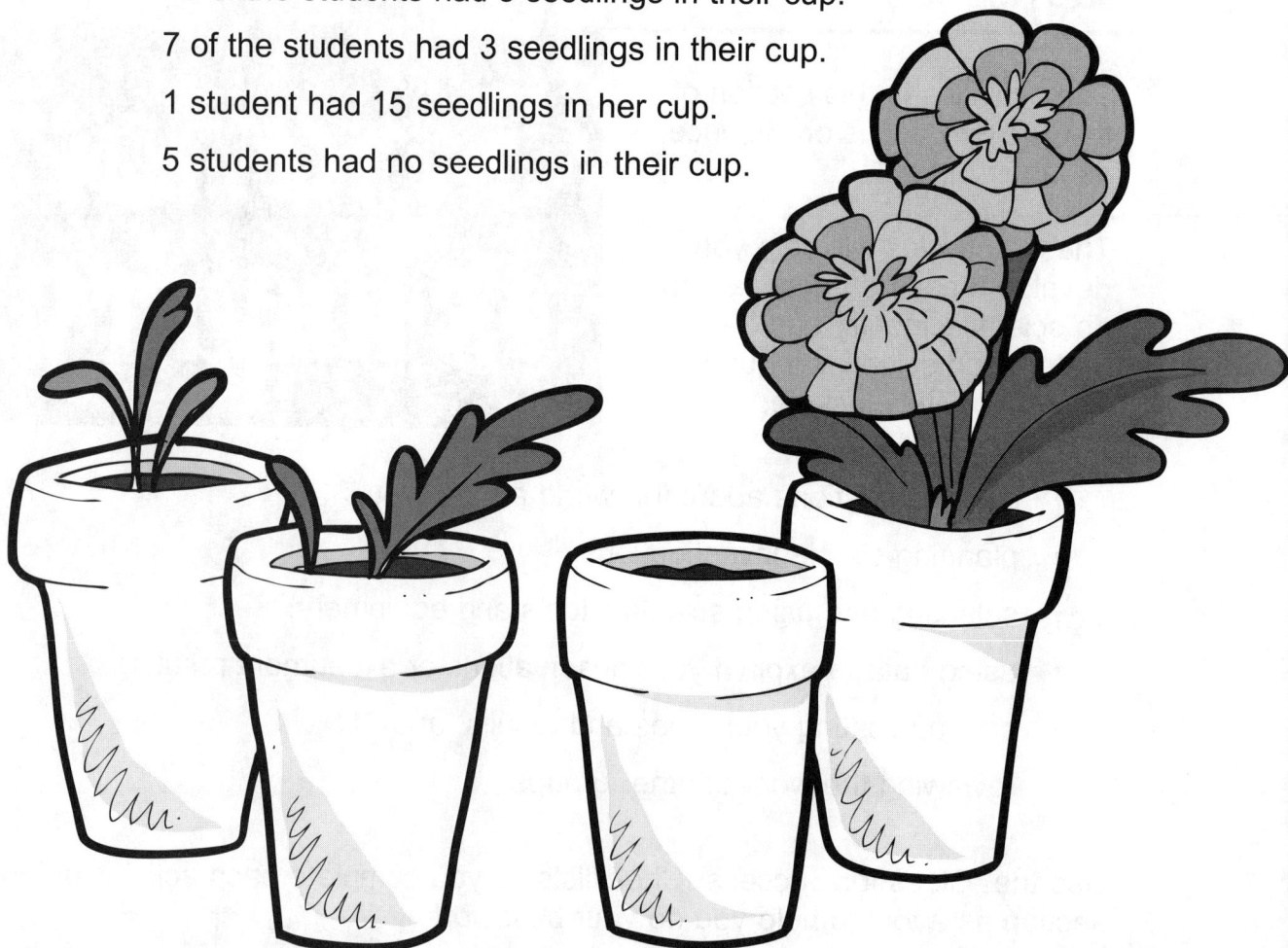

Step 2

Complete the Checklist "Clues for Success."
The checklist will help you to read and think like a scientist.

Clues for Success

☐ **C**arefully read the information.

☐ **L**ook at any illustrations or diagrams.
They may provide you with additional information to answer the question.

☐ **U**nderstand the way you are asked to answer the question.
   ☐ Graph
   ☐ Chart
   ☐ Diagram
   ☐ Complete sentences
   ☐ Phrase
   ☐ Filled circle

☐ **E**xamine the information given.
   ☐ Reread the questions.
   ☐ Underline key words or phrases.
   ☐ Think about what the questions are asking.

☐ **S**ee if your answers match the questions.

Step 3

Use the information from "The Marigold Planting Project" to complete the graphic organizer.

Make a bar graph to show the results of the students' marigold planting experience. Be sure to title your graph.

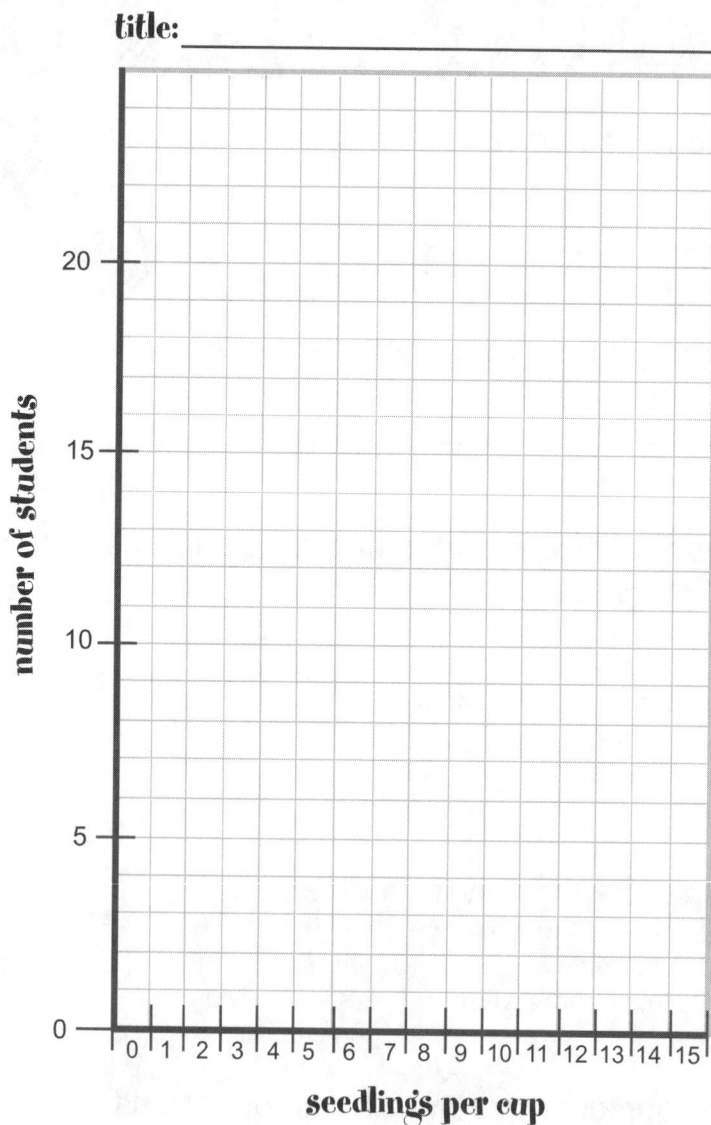

title: _____

number of students

20 —

15 —

10 —

5 —

0 —

0 1 2 3 4 5 6 7 8 9 10 11 12 13 14 15

seedlings per cup

Step 4

Answer the following question for "The Marigold Planting Project" using information from your graphic organizer.

1. Create a pictograph of the results of the students' marigold planting experience. Be sure to explain the symbols you use for the cups and seedlings.

Step 1

Read the scenario "Measuring Up."

Measuring Up

The boys and girls were having a discussion about the size of their hands. Both groups thought their hands were the largest. Their teacher wanted the students to compare their hands using scientific evidence as proof about the size. She also told them they had to agree on what measurement to use.

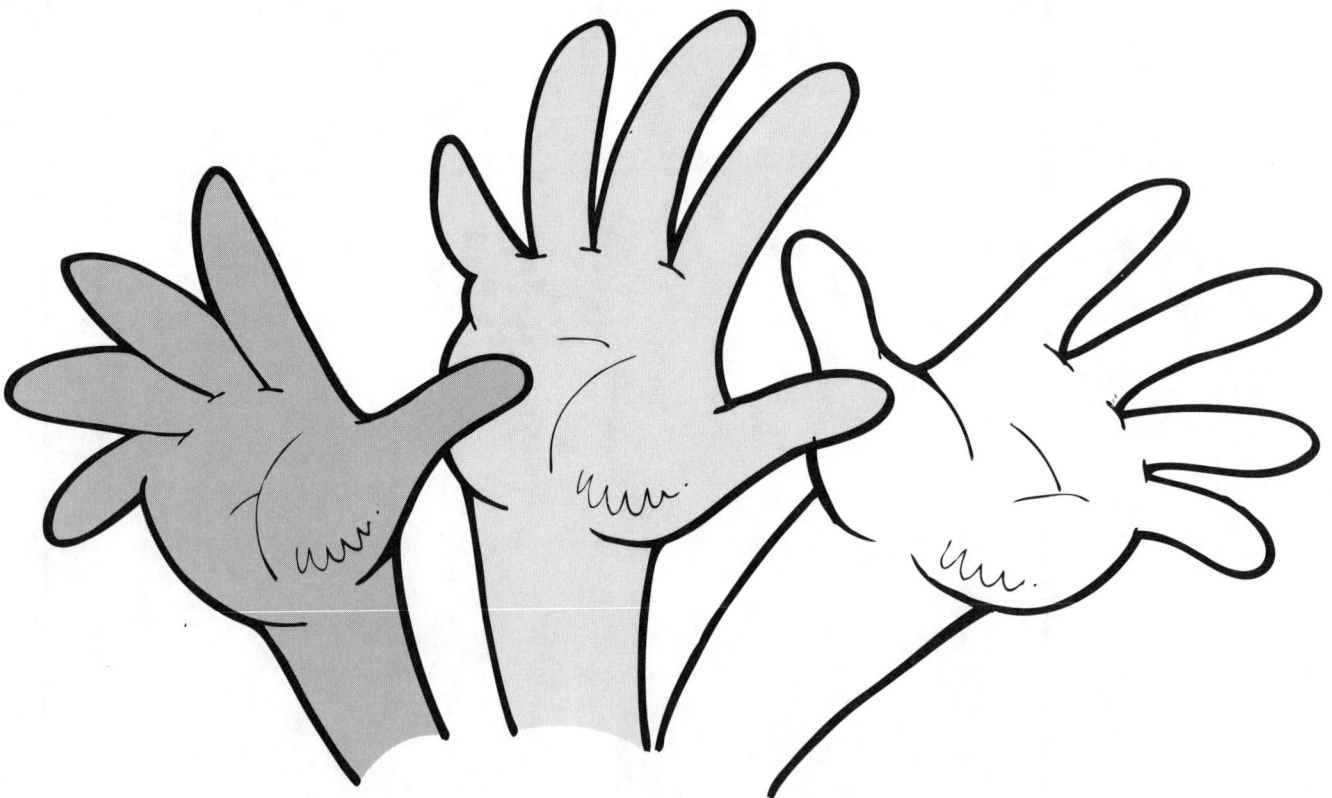

      © Englefield & Associates, Inc.

Step 2

Complete the Checklist "Clues for Success."

The checklist will help you to read and think like a scientist.

Clues for Success

☐ **C**arefully read the information.

☐ **L**ook at any illustrations or diagrams.
   They may provide you with additional information to answer the question.

☐ **U**nderstand the way you are asked to answer the question.
   - ☐ Graph
   - ☐ Chart
   - ☐ Diagram
   - ☐ Complete sentences
   - ☐ Phrase
   - ☐ Filled circle

☐ **E**xamine the information given.
   - ☐ Reread the questions.
   - ☐ Underline key words or phrases.
   - ☐ Think about what the questions are asking.

☐ **S**ee if your answers match the questions.

Activity 2

Step

3

Use the information from "Measuring Up" to complete the graphic organizer.

On the drawing of the hand, use a line and arrows to show what area you would measure for the comparison.

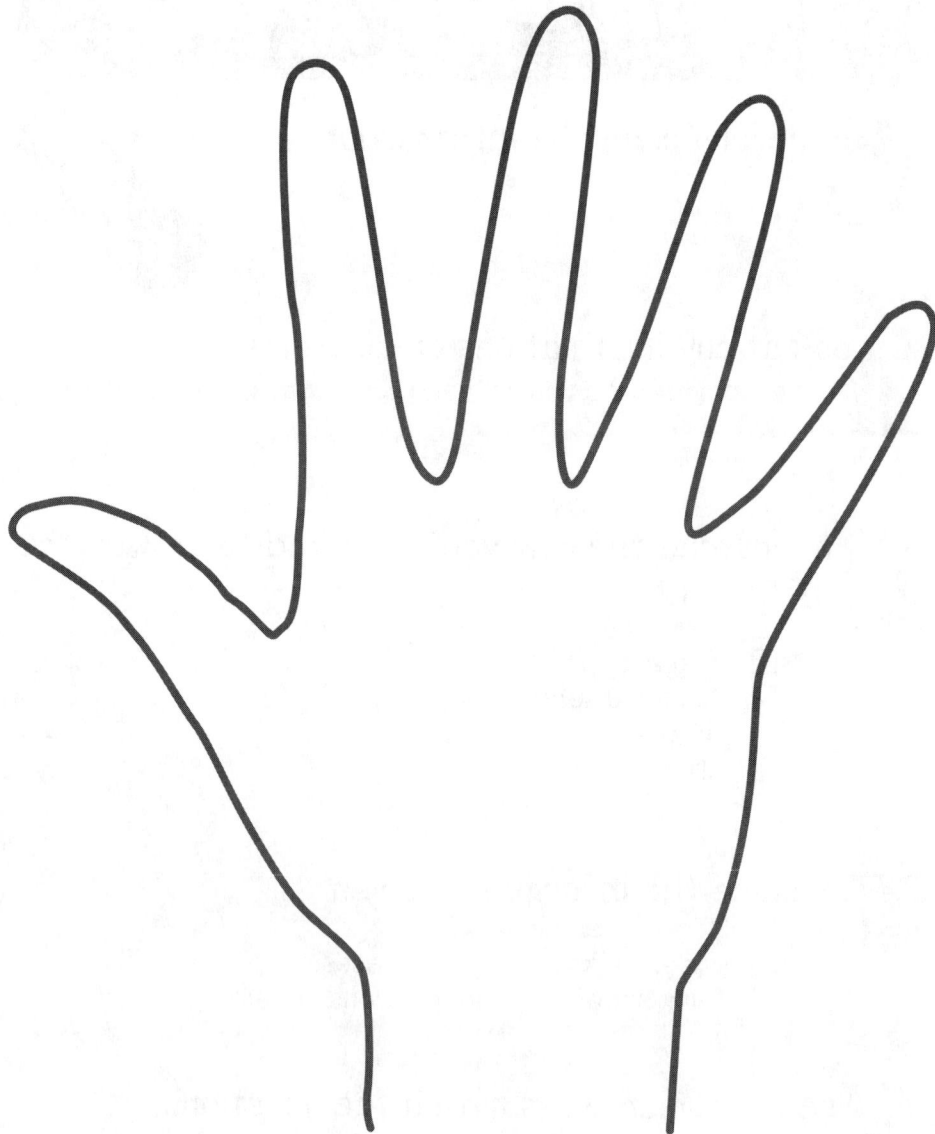

Step 4

Answer the following questions for "Measuring Up" using information from your graphic organizer.

1. What measurement would you suggest to convince the teacher?

   ○ wrist to fingertip   ○ width of hand   ○ other

   _____

2. What information would you use to convince the teacher to select your choice?

   _____

   _____

   _____

3. What tools would you use to **make** and **record** your measurements?

   _____

   _____

   _____

   _____

Step 1

Read the scenario "All About Worms."

All About Worms

The students researched earthworms on the Internet. They made a list of the information they learned and hung it on the bulletin board.

Some Facts We Learned About Earthworms

1. Earthworms come out of the ground when it rains.
2. Earthworms are food for birds.
3. Earthworms may be a few inches to a few feet long.
4. Earthworms aerate (make holes in) the soil.
5. Earthworms enrich the soil.
6. The young emerge as small, fully formed earthworms.
7. Earthworms live underground.
8. Earthworms decompose leaves, fruit, and fungi.
9. Earthworms might regenerate or replace lost segments.
10. Earthworms are also known as angleworms, rainworms, and night crawlers.
11. Earthworms will dry out without moisture.

Students continued learning about earthworms by making an earthworm home out of a 2-liter bottle. The class listed the steps to make the earthworm home on the board.

Steps to Create an Earthworm Home

1. Cut off the top of the bottle.

2. Put soil and sand in the bottle.

3. Add pieces of apple, grapes, and shredded paper.

4. Spray the soil in the bottle with water.

5. Add more soil.

6. Carefully add three worms.

7. Spray inside the bottle with more water.

8. Cover the bottle with black construction paper.

9. Replace the top of the bottle.

Step 2

Complete the Checklist "Clues for Success."

The checklist will help you to read and think like a scientist.

Clues for Success

☐ **C**arefully read the information.

☐ **L**ook at any illustrations or diagrams.
   They may provide you with additional information to answer the question.

☐ **U**nderstand the way you are asked to answer the question.
   - ☐ Graph
   - ☐ Chart
   - ☐ Diagram
   - ☐ Complete sentences
   - ☐ Phrase
   - ☐ Filled circle

☐ **E**xamine the information given.
   - ☐ Reread the questions.
   - ☐ Underline key words or phrases.
   - ☐ Think about what the questions are asking.

☐ **S**ee if your answers match the questions.

Step 3

Use the information from "All About Worms" to complete the graphic organizer.

Use the steps listed in the scenario to label the diagram below.

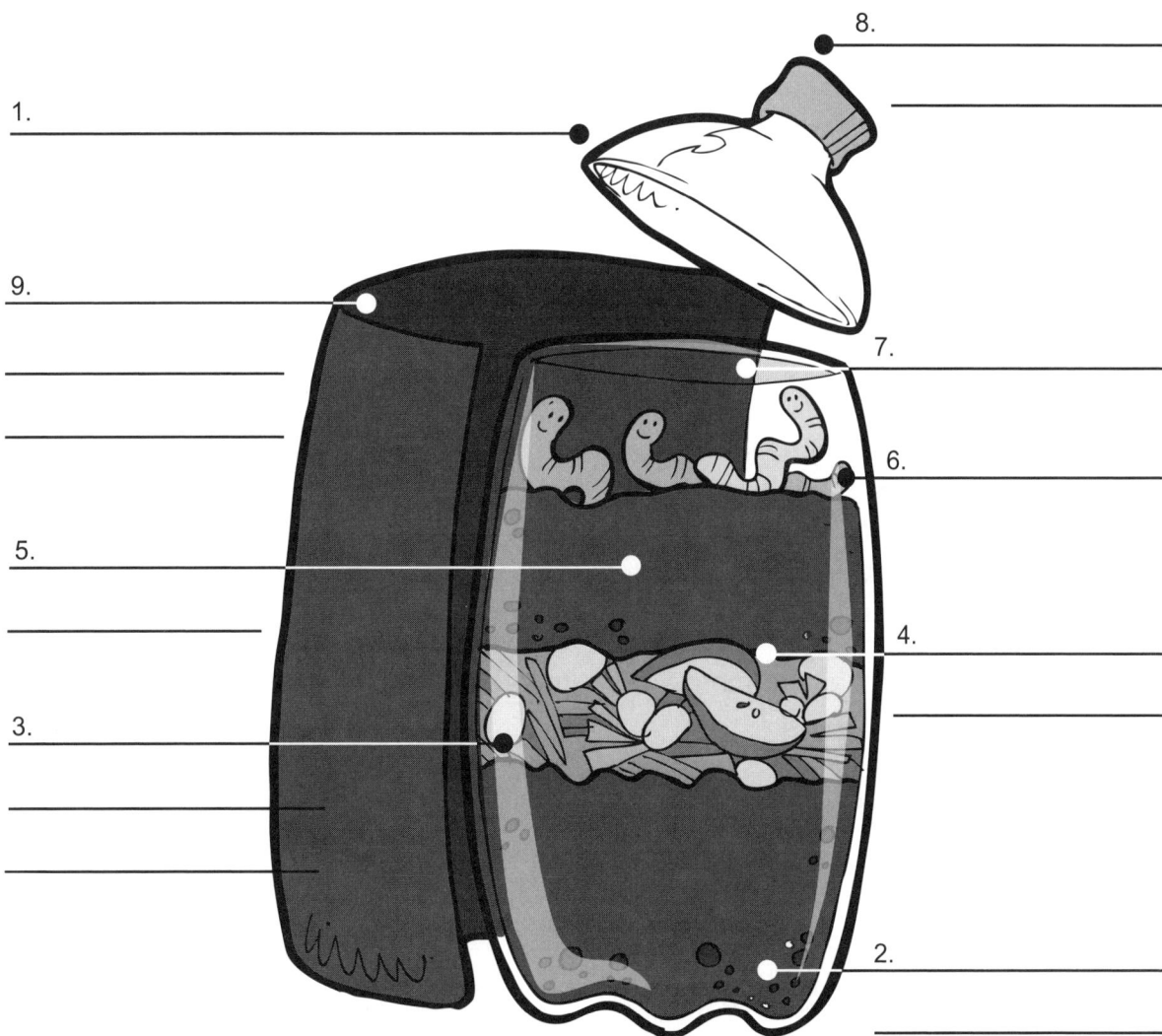

1. _____

8. _____

_____

9. _____

_____

7. _____

6. _____

5. _____

4. _____

_____

3. _____

_____

_____

2. _____

_____

Step 4

Answer the following question for "All About Worms" using information from your graphic organizer.

1. Read the list and use your graphic organizer to see the steps the students used to create their earthworm homes. Select the information from the poster that gives the reason for the step in the earthworm homes.

The students added pieces of apple, grapes, and shredded paper because

The students added more soil because

The students sprayed more water into the bottle because

The students covered the bottle with black construction paper because

Read the scenario
"Molecule Movement."

Molecule Movement

The students were learning that the molecules in hot water had more energy than the molecules in cold water. Their teacher wanted them to use this information when they made observations and recorded them, and then to explain their observations.

The teacher conducted a demonstration showing the movement of food coloring in jars of cold and hot water. The drawing of the students' observations is shown here.

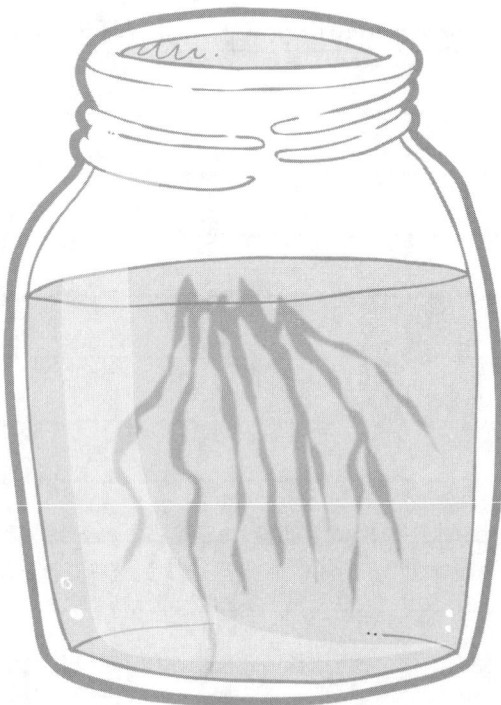

Jar A   Jar B

Step 2

Complete the Checklist "Clues for Success."

The checklist will help you to read and think like a scientist.

Clues for Success

☐ **C**arefully read the information.

☐ **L**ook at any illustrations or diagrams.
They may provide you with additional information to answer the question.

☐ **U**nderstand the way you are asked to answer the question.
☐ Graph
☐ Chart
☐ Diagram
☐ Complete sentences
☐ Phrase
☐ Filled circle

☐ **E**xamine the information given.
☐ Reread the questions.
☐ Underline key words or phrases.
☐ Think about what the questions are asking.

☐ **S**ee if your answers match the questions.

Step 3

Use the information from "Molecule Movement" to complete the graphic organizer.

Draw the observations the class made in the two jars. Be sure to label the jars A or B, as well as labeling them as cold water or hot water.

Step 4

Answer the following question for "Molecule Movement" using information from your graphic organizer.

1. What do the illustrations of Jar A and Jar B show about the movement of the food coloring in the jars of cold and hot water? Use the information you know about temperature to help you explain the movement of the food coloring in each jar.

hot water _____

_____

_____

_____

_____

cold water _____

_____

_____

_____

Step 1

Read the scenario
"Seed Dispersal."

Seed Dispersal

Most plants make seeds as a part of their reproduction cycle. These seeds are dispersed (travel away) from the plants to get enough water, soil, and sunlight for their survival. The seeds travel in many ways and have adaptations (special features) that help them on their journey. The wind carries seeds that have wing-like shapes or seeds that are light and fluffy. Some seeds float away in moving water. Animals carry seeds that are covered with pointy spikes. Some plants pop open and send their seeds flying away from them. Birds and animals can carry seeds to new locations and scatter seeds in their droppings. Even people can carry seeds to new areas on the bottoms of their shoes.

 © Englefield & Associates, Inc.

Step 2

Complete the Checklist "Clues for Success."

The checklist will help you to read and think like a scientist.

Clues for Success

☐ **C**arefully read the information.

☐ **L**ook at any illustrations or diagrams.
    They may provide you with additional information to answer the question.

☐ **U**nderstand the way you are asked to answer the question.
    ☐ Graph
    ☐ Chart
    ☐ Diagram
    ☐ Complete sentences
    ☐ Phrase
    ☐ Filled circle

☐ **E**xamine the information given.
    ☐ Reread the questions.
    ☐ Underline key words or phrases.
    ☐ Think about what the questions are asking.

☐ **S**ee if your answers match the questions.

Step 3

Use the information from "Seed Dispersal" to complete the graphic organizer.

Reread the information in "Seed Dispersal" and make a list of key words to tell the ways seeds may be dispersed.

Key Words for Seed Dispersal

1. _____

_____

2. _____

_____

3. _____

_____

4. _____

_____

Activity 5

Answer the following question for "Seed Dispersal" using information from your graphic organizer.

1. Look at the seeds pictured. Then use a complete sentence to describe how it might be dispersed.

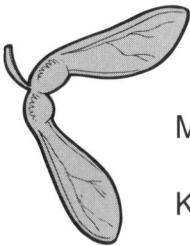

Maple seed

Key words: _____

_____

_____

Burdock

Key words: _____

_____

_____

Sweet Pea

Key words: _____

_____

_____

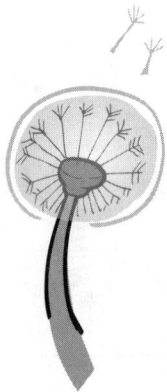

Dandelion

Key words: _____

_____

_____

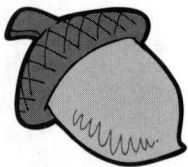

Acorn

Key words: _____

_____

_____

Step

1

Read the scenario "Forces and Motion."

Forces and Motion

The class was investigating forces and motion. Each of the three groups stacked a tower of six Science books. Then they leaned a cardboard tube with an open end four inches from the base of the books.

The students tested the distance a marble would roll as it exited the tube onto various surfaces including the carpet, a piece of cardboard, a rubber floor mat that has spikes to remove mud from shoes, and the table top.

The groups measured the following distances.

Group 1

carpet	4 inches
cardboard	8 inches

a floor mat with rubber spikes:
        stopped at bottom of tube

table top:
        rolled off of the table

Group 2

carpet	5 inches
cardboard	7 inches

a floor mat with rubber spikes:
        stopped at bottom of tube

table top:
        rolled 60 inches then fell off table

Group 3

carpet	4.5 inches
cardboard	8 inches

a floor mat with rubber spikes:
        stopped at bottom of tube

table top:
        rolled off of the table

Step 2

Complete the Checklist "Clues for Success."

The checklist will help you to read and think like a scientist.

Clues for Success

☐ **C**arefully read the information.

☐ **L**ook at any illustrations or diagrams.
     They may provide you with additional information to answer the question.

☐ **U**nderstand the way you are asked to answer the question.
     ☐ Graph
     ☐ Chart
     ☐ Diagram
     ☐ Complete sentences
     ☐ Phrase
     ☐ Filled circle

☐ **E**xamine the information given.
     ☐ Reread the questions.
     ☐ Underline key words or phrases.
     ☐ Think about what the questions are asking.

☐ **S**ee if your answers match the questions.

Step 3

Use the information from "Forces and Motion" to complete the graphic organizer.

Complete the data table showing the groups' results for the distance rolled on the carpet, cardboard, rubber mat, and tabletop.

group	distance rolled in inches			
	carpet	cardboard	rubber mat	table top
1				
2				
3				

Activity 6

Step 4

Answer the following questions for "Forces and Motion" using information from your graphic organizer.

1. Use the information provided in the marble experiment to show the surfaces that were least resistant (easiest) and most resistant (difficult) to the marble rolling. Provide evidence from the graphic organizer.

least resistance	evidence

most resistance	evidence

2. Based on this experiment, do you think the marble would have more or less resistance on a cement floor than on a table top?

_____

_____

_____

Step 1

Read the scenario "Experiment Challenge."

Experiment Challenge

The equipment box on each table in the science lab had the following items:

cotton fabrics

safety glasses

timer

colored paper

index cards

magnet

thermometer

markers

calculator

magnifying glass

ruler

wool fabrics

The teams of students were challenged to design an experiment that would answer the question:

How does color affect temperature?

Each team of students would have the same amount of time to design an experiment.

Step 2

Complete the Checklist "Clues for Success."

The checklist will help you to read and think like a scientist.

Clues for Success

☐ **C**arefully read the information.

☐ **L**ook at any illustrations or diagrams.
They may provide you with additional information to answer the question.

☐ **U**nderstand the way you are asked to answer the question.
- ☐ Graph
- ☐ Chart
- ☐ Diagram
- ☐ Complete sentences
- ☐ Phrase
- ☐ Filled circle

☐ **E**xamine the information given.
- ☐ Reread the questions.
- ☐ Underline key words or phrases.
- ☐ Think about what the questions are asking.

☐ **S**ee if your answers match the questions.

Step 3

Use the information from "Experiment Challenge" to complete the graphic organizer.

Tommy's team decided they would compare white and purple colored fabrics for their experiment. They would place the fabrics next to each other in a sunny window.

These are the tools selected from the equipment box. Explain how they could use each of them.

item	reason
white cotton fabric	
purple cotton fabric	
thermometer	
index cards	
markers	

Step 4

Answer the following question for "Experiment Challenge" using information from your graphic organizer.

1. For their experiment Tommy's team cut equal pieces of white and purple cotton fabric. They wrapped the whole thermometer in the fabrics and placed them in a sunny window. After 2 hours they planned to record the temperatures on each thermometer.

Create a flow chart of the steps in Tommy's team's experiment.

Problem: How Does Color Affect Temperature?

Experiment: Compare the Temperature of Cotton Fabrics

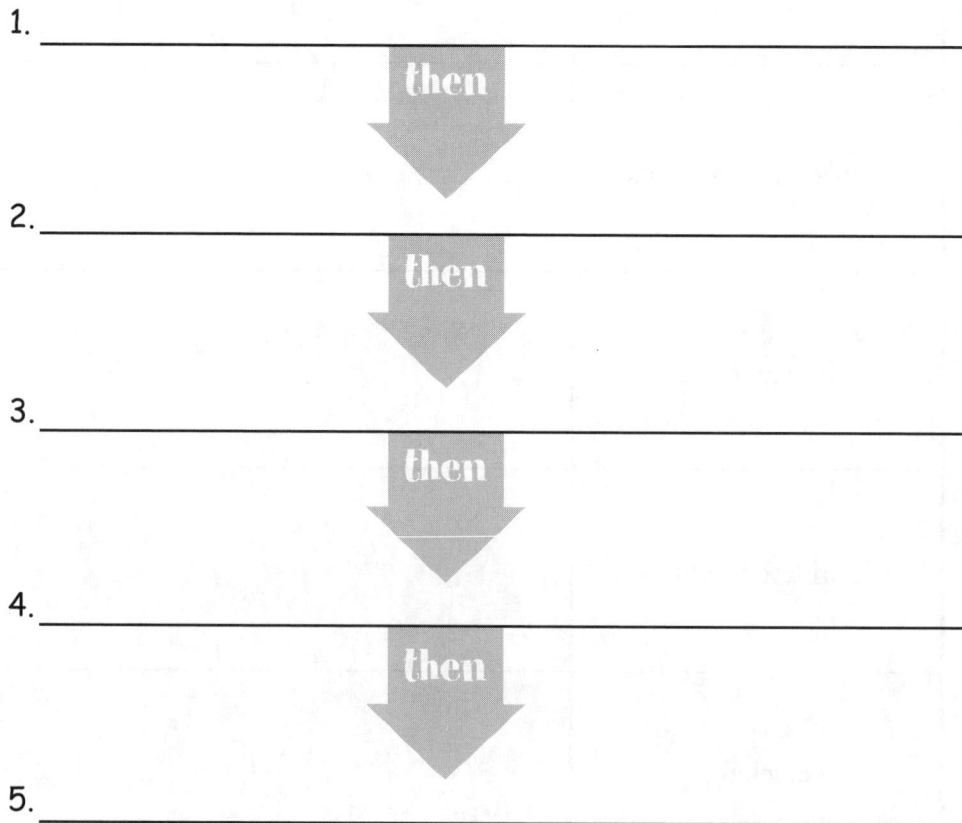

1. _____

then

2. _____

then

3. _____

then

4. _____

then

5. _____

 　　© Englefield & Associates, Inc.

Step 1

Read the scenario "Fruit Facts."

Fruit Facts

The students learned that fruit is more than something good to eat. In fact, some of their ideas about what fruit is changed with the students' new information. They learned that fruit is produced as a way for a plant to protect its seeds.

One or more seeds may be produced in the ovary of a flower when pollen is carried from one flower to another. The ovary is the part of the plant where the seeds grow and mature. As the seed develops, the ovary grows. This fleshy part that grows around the seed is known as the fruit.

Under the right living conditions, the fruit will protect the seed and it will grow into a plant. Other seeds will be eaten by organisms such as animals, birds, and people.

Step 2

Complete the Checklist "Clues for Success."

The checklist will help you to read and think like a scientist.

Clues for Success

☐ **C**arefully read the information.

☐ **L**ook at any illustrations or diagrams.
     They may provide you with additional information to answer the question.

☐ **U**nderstand the way you are asked to answer the question.
     ☐ Graph
     ☐ Chart
     ☐ Diagram
     ☐ Complete sentences
     ☐ Phrase
     ☐ Filled circle

☐ **E**xamine the information given.
     ☐ Reread the questions.
     ☐ Underline key words or phrases.
     ☐ Think about what the questions are asking.

☐ **S**ee if your answers match the questions.

Step 3

Use the information from "Fruit Facts" to complete the graphic organizer.

Using the information you know about fruit and the information provided, complete the "yes" or "no" checklist to determine if a food is a fruit. Read each statement then fill in your circle completely.

	yes	no
1. A fleshy part grows around the seed.	○	○
2. All seeds turn into fruit.	○	○
3. As the seed develops the ovary grows.	○	○

Step 4

Answer the following question for "Fruit Facts" using information from your graphic organizer.

1. Read each of the foods listed below. Decide if the food is a fruit based on the information you learned and put an X into the correct box.

	Fruit	Not a Fruit
Watermelon		
Tomato		
Carrot		
Cabbage		
Orange		
Apple		
Celery		
Pear		
Banana		

       © Englefield & Associates, Inc.

Chapter 2

The activities in this section of the book will focus on Physical Science.

These activities will help you investigate ideas about objects and materials that include:

- observing and describing their physical properties,

- describing their position and changes of position, and

- examining ideas about light, heat, electricity, and magnetism.

Use the "Clues for Success" Checklists as you complete each activity in this section as a tool to help you do your best work.

Step 1

Read the scenario "Classroom Objects."

Classroom Objects

The fourth graders are making a list of objects they found in their classroom. They noticed that they could group the objects based on several properties. Some of these included:

- The size that they could measure with a ruler or other measuring tool. The weight that could be measured on a scale.

- The temperature that they could measure with a thermometer.

- The volume of small objects that they would find by using a graduated cylinder.

- They could use their senses to determine color, texture, odor, and even taste with the permission of their teacher.

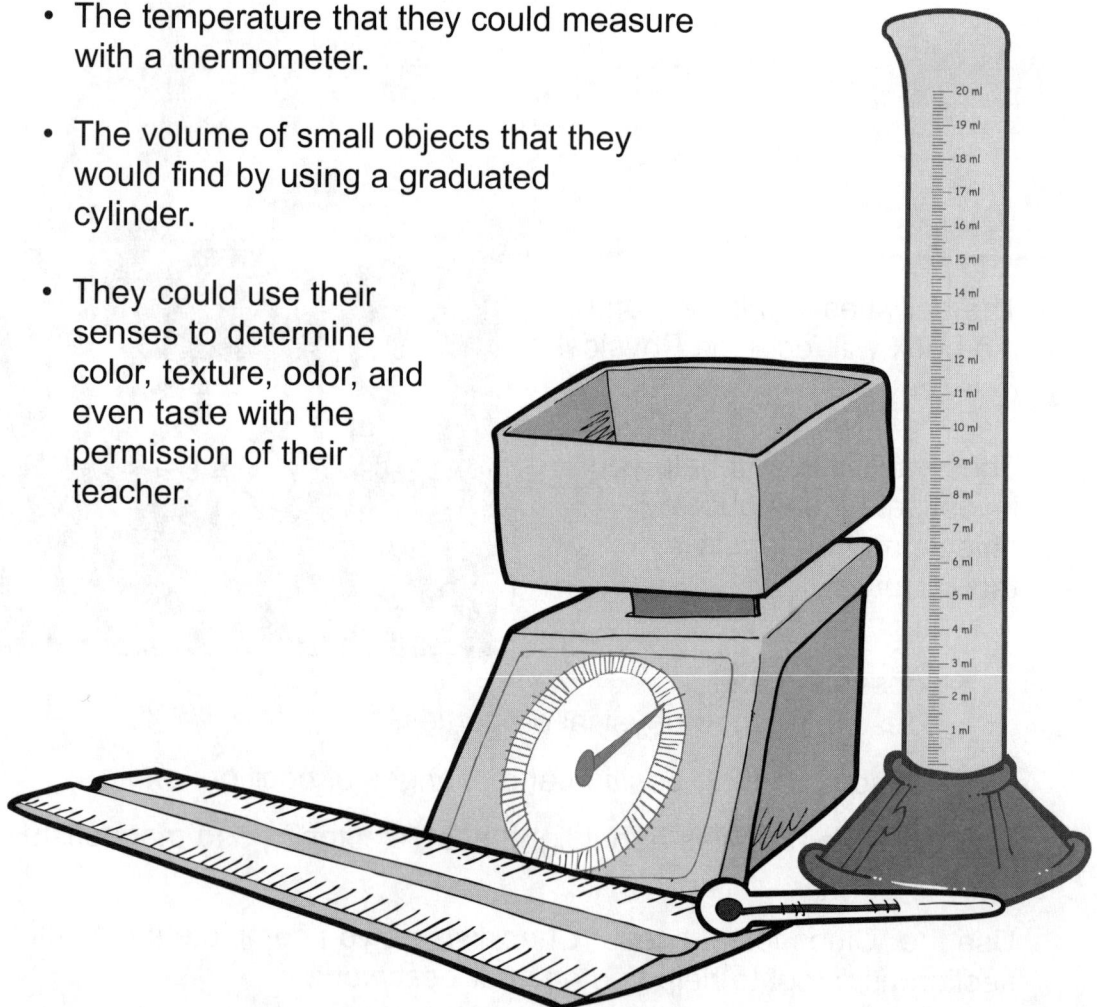

Step 2

Complete the Checklist "Clues for Success."
The checklist will help you to read and think like a scientist.

Clues for Success

☐ **C**arefully read the information.

☐ **L**ook at any illustrations or diagrams.
They may provide you with additional information to answer the question.

☐ **U**nderstand the way you are asked to answer the question.
- ☐ Graph
- ☐ Chart
- ☐ Diagram
- ☐ Complete sentences
- ☐ Phrase
- ☐ Filled circle

☐ **E**xamine the information given.
- ☐ Reread the questions.
- ☐ Underline key words or phrases.
- ☐ Think about what the questions are asking.

☐ **S**ee if your answers match the questions.

Step 3

Use the information from
"Classroom Objects"
to complete the graphic organizer.

Each object is pictured with a list of tools. On the lines provided tell which property can be observed or measured with each tool.

leaf

ruler _____

_____

eye _____

_____

scale _____

_____

magnifying glass _____

_____

thermometer _____

_____

beaker
of water

eye _____

_____

scale _____

_____

graduated cylinder _____

_____

Step 4

Answer the following question based on "Classroom Objects."

1. Using only your senses, what information can you find out about a pine cone?

pine cone

Hearing _____

_____

_____

_____

Feeling _____

_____

_____

_____

Smelling _____

_____

_____

_____

Seeing _____

_____

_____

_____

Step 1

Read the scenario "Mix and Separate."

Mix and Separate

Martin's class was learning about separating mixtures. Each group of students received a plastic cup filled with assorted materials. The cup included sand, rocks, popcorn kernels, iron filings, marbles, and paper clips.

On each group's tray they also received:
- a spoon
- a cup of water
- 2 squares of plastic mesh (one with small holes 1/8 inch and one with large holes 1/4 inch), and
- a magnet.

Step 2

Complete the Checklist "Clues for Success."

The checklist will help you to read and think like a scientist.

Clues for Success

☐ **C**arefully read the information.

☐ **L**ook at any illustrations or diagrams.
They may provide you with additional information to answer the question.

☐ **U**nderstand the way you are asked to answer the question.
- ☐ Graph
- ☐ Chart
- ☐ Diagram
- ☐ Complete sentences
- ☐ Phrase
- ☐ Filled circle

☐ **E**xamine the information given.
- ☐ Reread the questions.
- ☐ Underline key words or phrases.
- ☐ Think about what the questions are asking.

☐ **S**ee if your answers match the questions.

Step 3

Use the information from "Mix and Separate" to complete the graphic organizer.

Complete the table to show what the students will separate from the mixture using the illustrated tools.

use	what will be separated
spoon	
plastic mesh with large holes	
plastic mesh with small holes	
magnet	

          © Englefield & Associates, Inc.

Step 4

Answer the following questions based on "Mix and Separate."

1. The students will also use their senses to help them sort out the mixtures. List two senses the students will use. Create a symbol for each of these senses.

Sense 1   Sense 2

_____   _____

Symbol 1   Symbol 2

2. Complete the table using your symbols to describe how the students will use them when they separate the mixture.

sense	how it will help separate the mixture

Step 1

Read the scenario
"Push and Pull."

Push and Pull

The fourth grade teacher often takes students outside to explore ideas in Science. Today she took the students to the playground to demonstrate the terms "push" and "pull."

The students played the game Tug of War. She explained that "pull" means to bring an object toward you and "push" means to move an object away from you. She also explained that pushes and pulls are forces. The forces can work together like the pulling of all students on one side of the rope. The forces can work against each other like the work of the students on the opposite side of the rope.

Step 2

Complete the Checklist "Clues for Success."

The checklist will help you to read and think like a scientist.

Clues for Success

☐ **C**arefully read the information.

☐ **L**ook at any illustrations or diagrams.
  They may provide you with additional information to answer the question.

☐ **U**nderstand the way you are asked to answer the question.
  ☐ Graph
  ☐ Chart
  ☐ Diagram
  ☐ Complete sentences
  ☐ Phrase
  ☐ Filled circle

☐ **E**xamine the information given.
  ☐ Reread the questions.
  ☐ Underline key words or phrases.
  ☐ Think about what the questions are asking.

☐ **S**ee if your answers match the questions.

Step 3

Use the information from "Push and Pull" to complete the graphic organizer.

Show the force of the pulling on the Tug of War games shown below. Draw a large arrow on the side of the rope that is pulled with greater force (stronger) and a small arrow on the side of the rope that is pulled with less force (weaker). If the force is equal, make the arrows equal size.

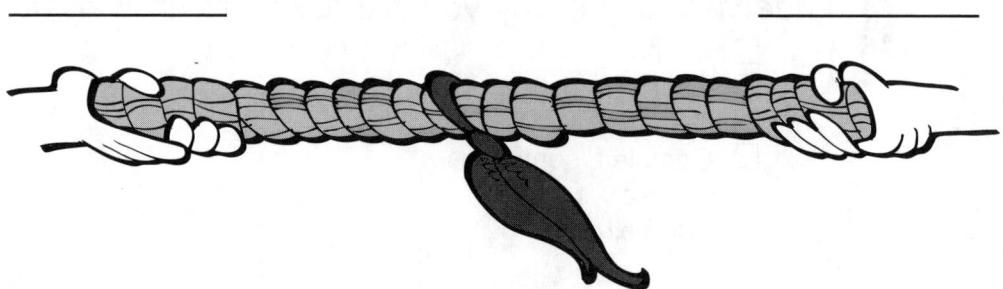

_____   _____

_____   _____

_____   _____

COPYING IS PROHIBITED   © Englefield & Associates, Inc.

Step 4

Answer the following question based on "Push and Pull."

1. The students on the playground then were asked to use a swing to demonstrate different kinds of "pushes." They are asked to show the difference between a strong and a weak push. Draw a picture to illustrate this difference and write a complete sentence to explain your answer.

_____

_____

Step 1

Read the scenario "Reflection Reflections."

Reflection Reflections

Mark's class is learning about mirrors and reflection in school. Mark has an idea about reflections because he saw his image in several objects each day—the mirror when he combed his hair and even in the toaster and in his spoon when he ate breakfast. So the next morning Mark really paid attention to his reflections.

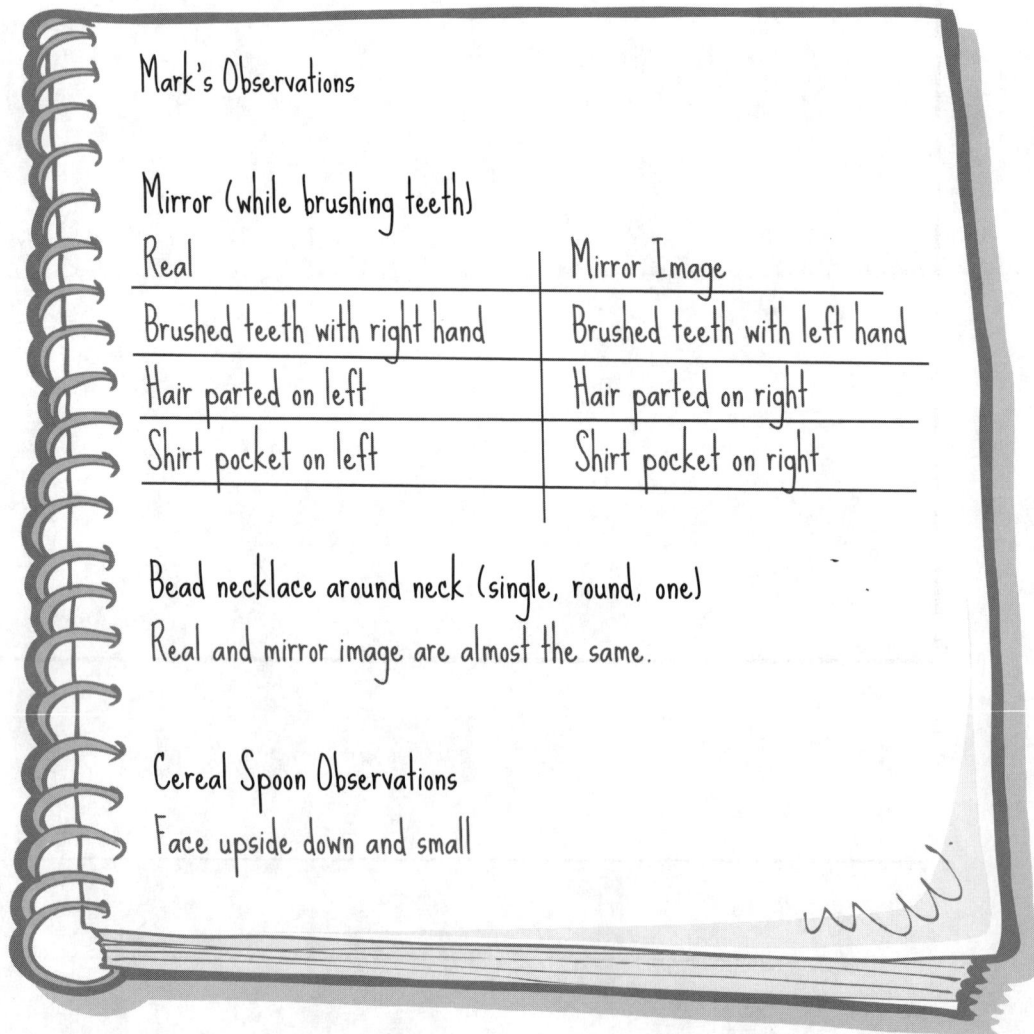

Mark's Observations

Mirror (while brushing teeth)

Real	Mirror Image
Brushed teeth with right hand	Brushed teeth with left hand
Hair parted on left	Hair parted on right
Shirt pocket on left	Shirt pocket on right

Bead necklace around neck (single, round, one)
Real and mirror image are almost the same.

Cereal Spoon Observations
Face upside down and small

 © Englefield & Associates, Inc.

Step 2

Complete the Checklist "Clues for Success."

The checklist will help you to read and think like a scientist.

Clues for Success

☐ **C**arefully read the information.

☐ **L**ook at any illustrations or diagrams.
    They may provide you with additional information to answer the question.

☐ **U**nderstand the way you are asked to answer the question.
- ☐ Graph
- ☐ Chart
- ☐ Diagram
- ☐ Complete sentences
- ☐ Phrase
- ☐ Filled circle

☐ **E**xamine the information given.
- ☐ Reread the questions.
- ☐ Underline key words or phrases.
- ☐ Think about what the questions are asking.

☐ **S**ee if your answers match the questions.

Step 3

Use the information from "Reflection Reflections" to complete the graphic organizer.

Mark's classmate John decided to record his own reflection. Look at John. Draw the reflection (mirror image) John would see when he looked in the mirror.

John

John's Reflection

In his journal, John wrote the real and mirror image are almost the same. What did he mean when he wrote that?

_____

_____

_____

Step 4

Answer the following questions based on "Reflection Reflections."

1. Complete the chart to show two pieces of information Mark observed about reflected images.

Reflected Images

Mirrors (Shiny flat surfaces)	Spoons (Shiny curved surfaces)

2. Mark wondered about the image of the bead necklace. Why do you think the image of the bead necklace stayed the same as the real one?

_____

_____

_____

Step 1

Read the information about heat production.

Heating is a condition that adds or increases energy. Heat can be produced in many ways.

Burning—igniting substances in a chemical reaction that uses up oxygen.

Rubbing—moving one thing back and forth against another.

Mixing—adding one substance to another.

     © Englefield & Associates, Inc.

Step 2

Complete the Checklist "Clues for Success."
The checklist will help you to read and think like a scientist.

Clues for Success

☐ **C**arefully read the information.

☐ **L**ook at any illustrations or diagrams.
They may provide you with additional information to answer the question.

☐ **U**nderstand the way you are asked to answer the question.
   ☐ Graph
   ☐ Chart
   ☐ Diagram
   ☐ Complete sentences
   ☐ Phrase
   ☐ Filled circle

☐ **E**xamine the information given.
   ☐ Reread the questions.
   ☐ Underline key words or phrases.
   ☐ Think about what the questions are asking.

☐ **S**ee if your answers match the questions.

Step 3

Use the information about heat production to complete the graphic organizer.

Use your past experiences with heat to help you complete the concept map about the different ways heat can be generated.

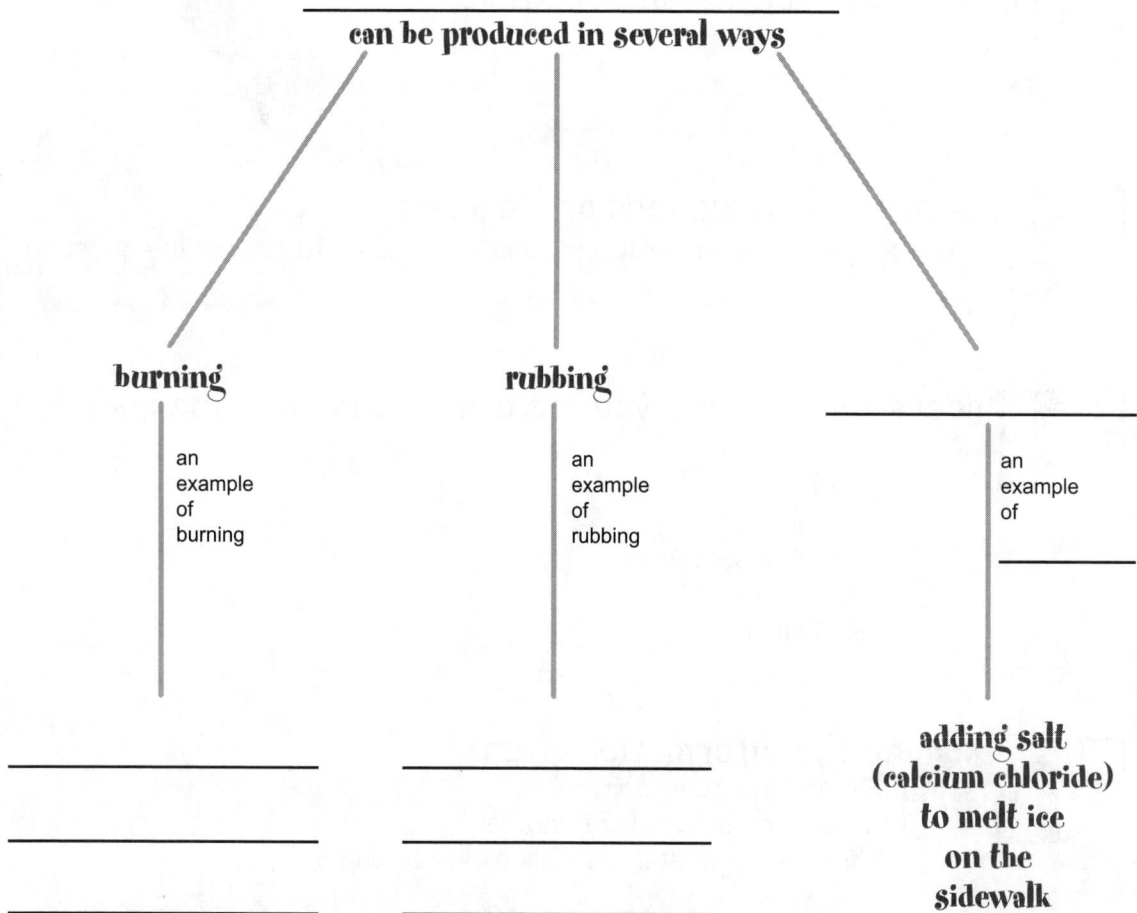

_____

can be produced in several ways

burning   **rubbing**   _____

an example of burning   an example of rubbing   an example of _____

adding salt (calcium chloride) to melt ice on the sidewalk

_____   _____

_____   _____

_____   _____

Step 4

Answer the following question based on information about heat production.

The class made a list of when they saw things burning. The list included:
  burning leaves
  a campfire
  forest fires
  using a match to light a candle

1. Do you think heat created by burning is always helpful?

   ○ yes   ○ no

   Give an example to support your answer.

   _____

   _____

   _____

   _____

Step 1

Read the information about electrical energy.

Electrical energy (electricity) is produced by moving electrons. Electrical energy is used in many ways in our daily lives, including the energy to light and heat our homes, schools, and buildings. Electricity is the force that is used in many homes to cook food. When a computer is used to research a school project or play video games, electricity is being used. Entertainment activities like watching television, listening to CDs, or even playing an electric guitar are all using electrical energy. An electric circuit is the path of electron flow that creates energy.

Step 2

Complete the Checklist "Clues for Success."

The checklist will help you to read and think like a scientist.

Clues for Success

☐ **C**arefully read the information.

☐ **L**ook at any illustrations or diagrams.
They may provide you with additional information to answer the question.

☐ **U**nderstand the way you are asked to answer the question.
   ☐ Graph
   ☐ Chart
   ☐ Diagram
   ☐ Complete sentences
   ☐ Phrase
   ☐ Filled circle

☐ **E**xamine the information given.
   ☐ Reread the questions.
   ☐ Underline key words or phrases.
   ☐ Think about what the questions are asking.

☐ **S**ee if your answers match the questions.

Activity 6

Step 3

Use the information about electrical energy to complete the graphic organizer.

Read each of the uses of electricity listed on the next page. Cut and paste each use into the appropriate category. Then add one use you can think of into each category.

Light	Heat	Sound

 © Englefield & Associates, Inc.

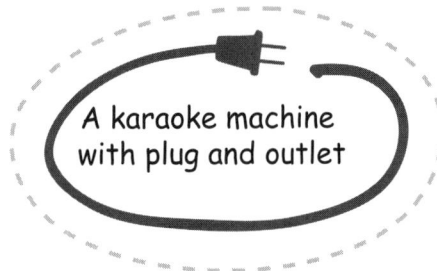

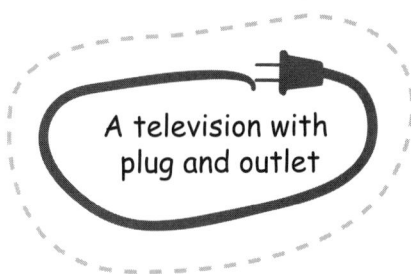

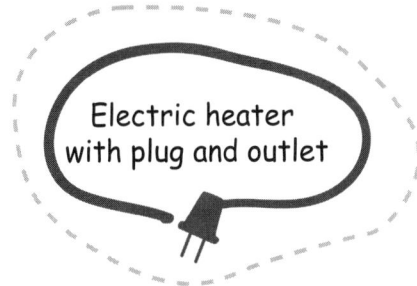

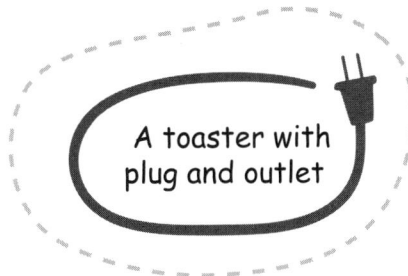

Step 4

Answer the following question based on the information about electrical energy.

1. Electricity must have a complete loop (circuit) to allow the movement of electrons to flow.

This is a series circuit.

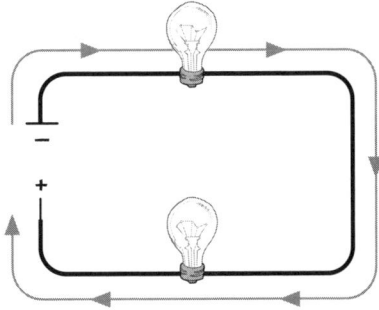

There is only one path for the electrons to flow and both bulbs are on the same path of the electron flow (electricity).

This is a parallel circuit.

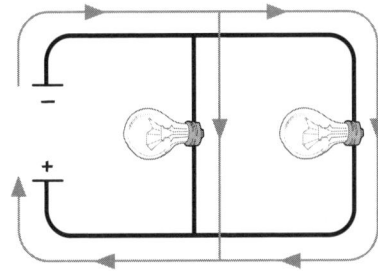

There are two paths for the electrons to flow. Each bulb has its own path of electron flow (electricity).

Look at this illustration. Draw arrows to show the flow of electricity.

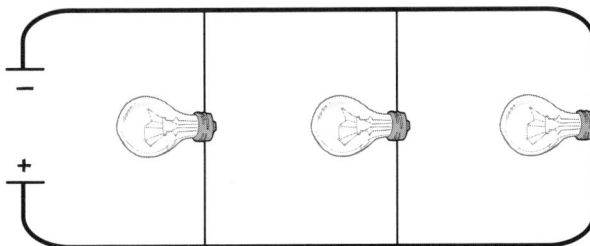

Is this a series or a parallel circuit? _____

Step 1

Read the scenario "Magnetic Attraction."

Magnetic Attraction

Margaret was playing with a magnet. Magnetic force is the force of attraction between magnets and magnetic objects. She made a checklist of objects that are attracted or not attracted to the magnet.

Object	Attracted	Not attracted
a nail	X	
a balloon		X
a paper clip	X	
a safety pin	X	
a sheet of paper		X
a marble		X

Step 2

Complete the Checklist "Clues for Success."
The checklist will help you to read and think like a scientist.

Clues for Success

☐ **C**arefully read the information.

☐ **L**ook at any illustrations or diagrams.
They may provide you with additional information to answer the question.

☐ **U**nderstand the way you are asked to answer the question.
  - ☐ Graph
  - ☐ Chart
  - ☐ Diagram
  - ☐ Complete sentences
  - ☐ Phrase
  - ☐ Filled circle

☐ **E**xamine the information given.
  - ☐ Reread the questions.
  - ☐ Underline key words or phrases.
  - ☐ Think about what the questions are asking.

☐ **S**ee if your answers match the questions.

Step 3

Use the information from "Magnetic Attraction" to complete the graphic organizer.

Use the information that Margaret discovered to construct a flow chart about objects that are attracted to magnets and objects that are not attracted to magnets. First, place the objects that Margaret tested in the appropriate place on the chart. Then, add two more objects that are attracted and not attracted to the magnet.

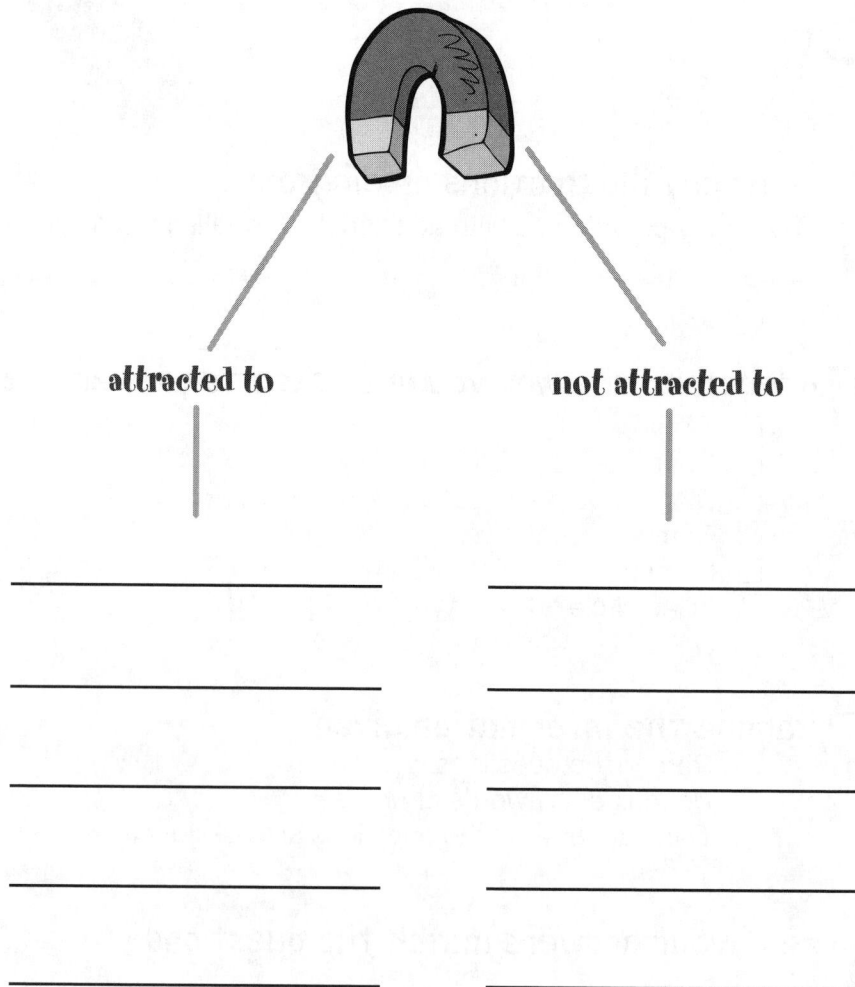

attracted to   **not attracted to**

_____   _____

_____   _____

_____   _____

_____   _____

_____   _____

Step 4

Answer the following questions based on "Magnetic Attraction."

1. Using the information that is on the flow chart, where would you place a brown-paper lunch bag on the chart? Use a completely filled circle to show your answer choice.

    ○ attracted to the magnet   ○ not attracted to the magnet

    Why did you make this choice?

    _____

    _____

    _____

2. Using the information that is on the flow chart, where would you place a tack on the flow chart? Use a completely filled circle to show your answer choice.

    ○ attracted to the magnet   ○ not attracted to the magnet

    Why did you make this choice?

    _____

    _____

    _____

3. How are all objects attracted to a magnet the same?

_____

_____

_____

_____

Chapter 3

The activities in this section of the book will focus on Life Science.

These activities will help you investigate:

- the characteristics of organisms and the way they meet their needs for life,

- the life cycles of organisms, and

- the ways organisms interact with their environment.

Use the "Clues for Success" Checklists as you complete each activity in this section as a tool to help you do your best work.

Step 1

Read the scenario
"Parts of a Whole."

Parts of a Whole

Every living organism, plant and animal, is comprised of different parts (structures). These structures help the organism to grow, survive, and reproduce. For example, in plants

- cells reproduce to help the plant grow and develop,

- roots grow deeper into the ground to anchor the plant in the ground, and

- in some cases, seeds may be produced to make new plants.

Step 2

Complete the Checklist "Clues for Success."

The checklist will help you to read and think like a scientist.

Clues for Success

☐ **C**arefully read the information.

☐ **L**ook at any illustrations or diagrams.
   They may provide you with additional information to answer the question.

☐ **U**nderstand the way you are asked to answer the question.
   ☐ Graph
   ☐ Chart
   ☐ Diagram
   ☐ Complete sentences
   ☐ Phrase
   ☐ Filled circle

☐ **E**xamine the information given.
   ☐ Reread the questions.
   ☐ Underline key words or phrases.
   ☐ Think about what the questions are asking.

☐ **S**ee if your answers match the questions.

Step 3

Use the information from "Parts of a Whole" to complete the graphic organizer.

Identify the parts of the tree shown, then describe how that part of a tree (structure) helps the life of the tree (function it serves).

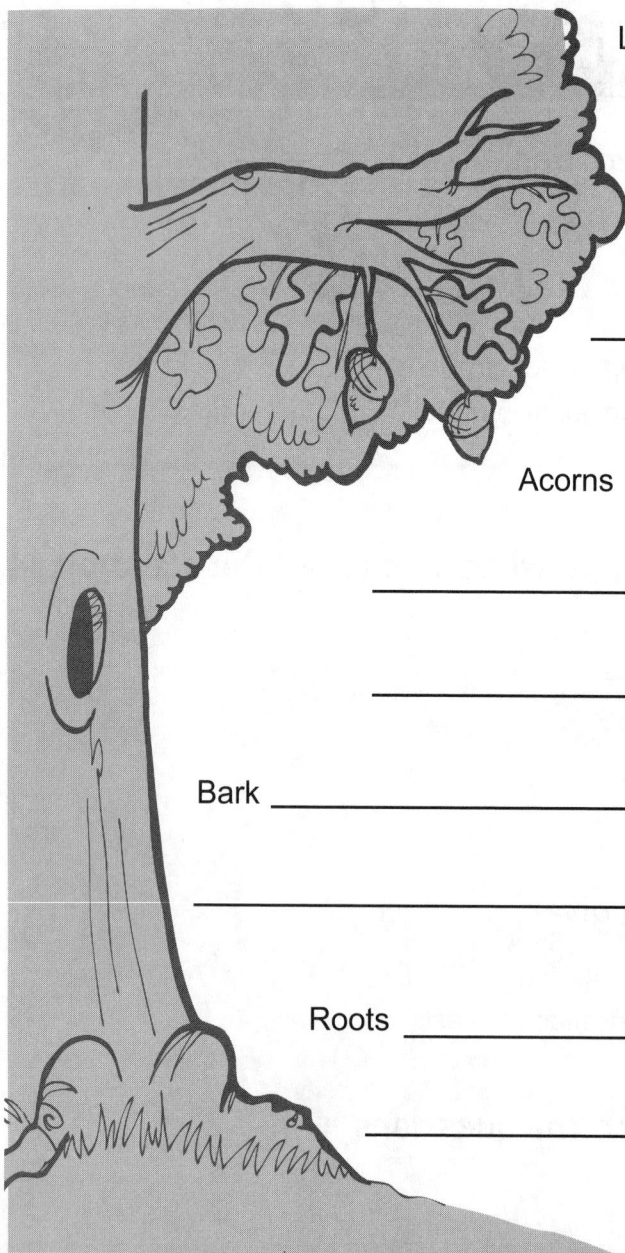

Leaves _____

_____

_____

_____

Acorns _____

_____

_____

Bark _____

_____

Roots _____

_____

Answer the following questions for "Parts of a Whole" using information from your graphic organizer.

1. Like other organisms, humans have body parts that help them in different ways. Two body parts that have different functions are arms and legs.

In what ways do humans use arms? Use complete sentences to describe three different functions.

1. _____

_____

2. _____

_____

3. _____

_____

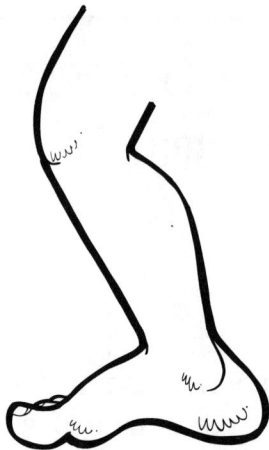

In what ways do humans use legs? Use complete sentences to describe three different functions.

1. _____

_____

2. _____

_____

3. _____

_____

Step 1

Read the scenario "Lots of Life Cycles."

Lots of Life Cycles

The students were learning that animals have life cycles that include being born, developing into adults, reproducing, and eventually dying. They had experience with butterflies and their transformation from egg to caterpillar to cocoon to adult. Today they wanted to learn more about other critters that they found outside their school.

The students prepared drawings from the information they found about the ant on the internet. The life cycle of the ant included:

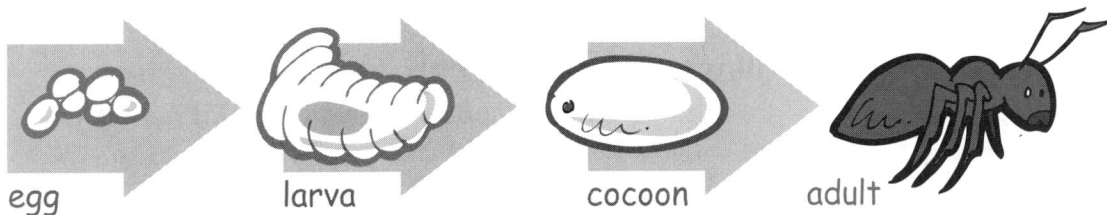

egg   larva   cocoon   adult

They looked up the life cycle of a spider in the encyclopedia and found out that the new baby spider looks like the adult only smaller.

egg   young   adult

They learned that the term for this change is metamorphosis. Metamorphosis is the transformation of the organism from birth to adult. The ant undergoes **complete metamorphosis** but the spider undergoes **incomplete metamorphosis**.

Step 2

Complete the Checklist "Clues for Success."

The checklist will help you to read and think like a scientist.

Clues for Success

☐ **C**arefully read the information.

☐ **L**ook at any illustrations or diagrams.
They may provide you with additional information to answer the question.

☐ **U**nderstand the way you are asked to answer the question.
- ☐ Graph
- ☐ Chart
- ☐ Diagram
- ☐ Complete sentences
- ☐ Phrase
- ☐ Filled circle

☐ **E**xamine the information given.
- ☐ Reread the questions.
- ☐ Underline key words or phrases.
- ☐ Think about what the questions are asking.

☐ **S**ee if your answers match the questions.

Step 3

Use the information from "Lots of Life Cycles" to complete the graphic organizer.

Draw and describe the four stages in the complete metamorphosis of the butterfly.

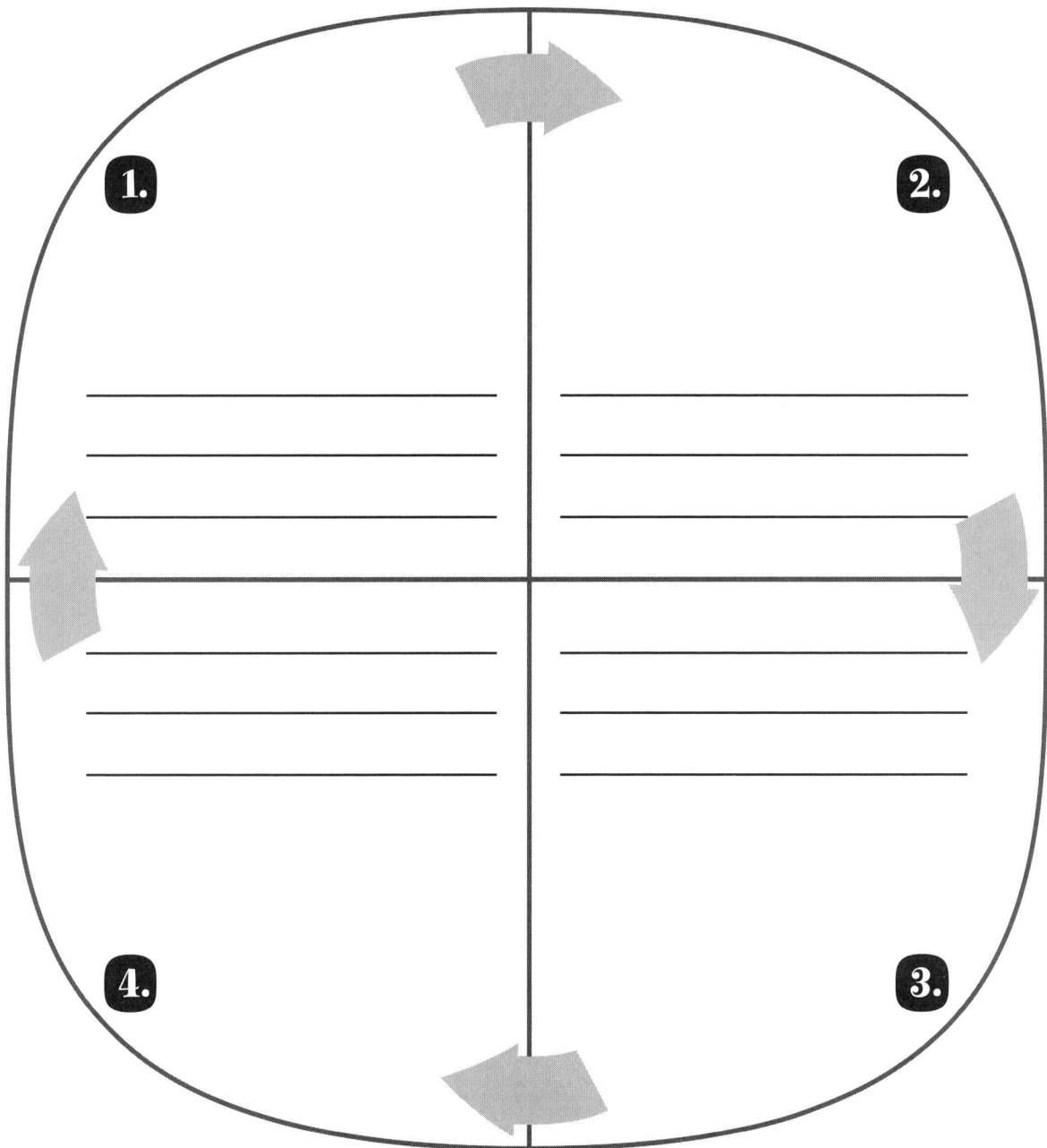

1.

2.

_____

_____

_____

_____

_____

_____

_____

_____

4.

3.

Step 4

Answer the following question for "Lots of Life Cycles" using information from your graphic organizer.

1. Create a Venn Diagram to compare the steps in the life cycle of the ant and the spider.

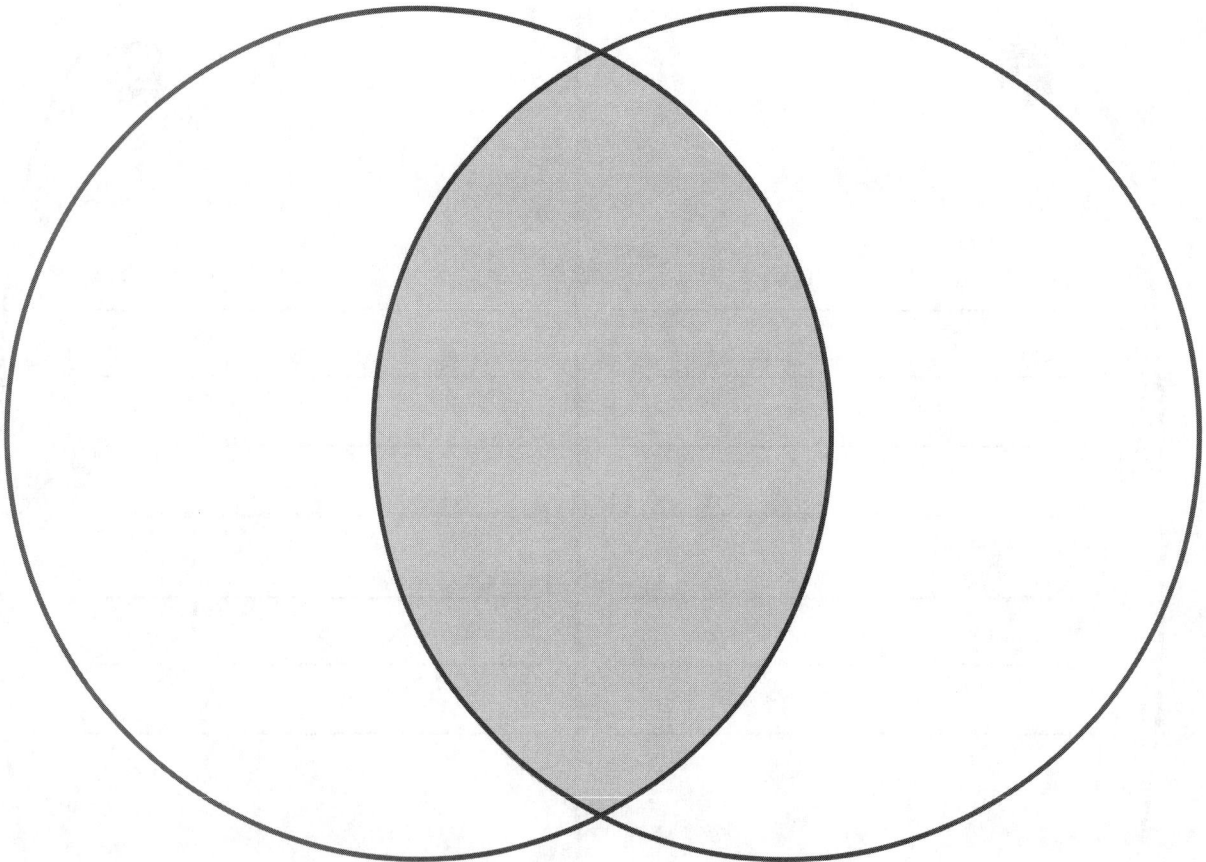

Step 1

Read the scenario
"Parents and their Young."

Parents and their Young

Young organisms like plants and animals generally have the same characteristics as their parents. Some of the ways parents and their young are alike include:

- their basic need for life such as their food source, habitat (home), and space needed to live,

- how they move (fly, swim, walk, etc.),

- their appearance (shape),

- the way they interact with their environment with sensory organs, like eyes and antennae; appendages, like wings, fins, or arms, and

- their job or function in the environment (niche).

Step 2

Complete the Checklist "Clues for Success."

The checklist will help you to read and think like a scientist.

Clues for Success

☐ **C**arefully read the information.

☐ **L**ook at any illustrations or diagrams.
They may provide you with additional information to answer the question.

☐ **U**nderstand the way you are asked to answer the question.
- ☐ Graph
- ☐ Chart
- ☐ Diagram
- ☐ Complete sentences
- ☐ Phrase
- ☐ Filled circle

☐ **E**xamine the information given.
- ☐ Reread the questions.
- ☐ Underline key words or phrases.
- ☐ Think about what the questions are asking.

☐ **S**ee if your answers match the questions.

Step 3

Use the information from
"Parents and their Young"
to complete the graphic organizer.

Select one of the baby animals or plant pictured and tell how the baby animal or plant is like its parents. Look at each characteristic listed on the chart and explain how the baby and parent are the same. Use a completely filled circle to show which adult organism and young organism you will be discussing.

appearance (how they look)	
movement (how they move)	
habitat (where they live)	

Activity 3

Step 4

Answer the following question for "Parents and their Young" using information from your graphic organizer.

1. You and your parents share many characteristics. On the chart below, list two ways you are different from your parents.

appearance (how they look)	1. 2.
job or function (niche)	1. 2.

Step 1

Read the scenario "Inherited and Learned Characteristics."

Inherited and Learned Characteristics

Many characteristics (traits) young organisms possess are inherited from their parents.

Some of these characteristics can be passed on from parent to young. They are inherited characteristics. They include the kind of hair they have, how tall they grow, and even the number of arms, legs, and eyes.

But some other features or abilities are the result of what the organisms experience in their environment. These changes in behavior cannot be passed on from parent to youth. They are learned characteristics like learning to ski, live in a new environment, or do a new job.

The combination of these inherited and learned characteristics help make organisms unique, but also similar to others.

Step 2

Complete the Checklist "Clues for Success."

The checklist will help you to read and think like a scientist.

Clues for Success

☐ **C**arefully read the information.

☐ **L**ook at any illustrations or diagrams.
     They may provide you with additional information to answer the question.

☐ **U**nderstand the way you are asked to answer the question.
    ☐ Graph
    ☐ Chart
    ☐ Diagram
    ☐ Complete sentences
    ☐ Phrase
    ☐ Filled circle

☐ **E**xamine the information given.
    ☐ Reread the questions.
    ☐ Underline key words or phrases.
    ☐ Think about what the questions are asking.

☐ **S**ee if your answers match the questions.

Step 3

Use the information from "Inherited and Learned Characteristics" to complete the graphic organizer.

Colleen made a list of her special features and talents. Help her sort the inherited characteristics (same as family members) and learned characteristics (things Colleen learned in her life).

> playing the piano
> brown eyes
> red hair
> playing goalie for my soccer team
> growing tall
> giving ice skating lessons at the park
> keeping my room tidy
> long fingers
> being good at doing magic tricks

Inherited Characteristics	Learned Characteristics

Step 4   Answer the following question for "Inherited and Learned Characteristics" using information from your graphic organizer.

1. Choose one of Colleen's learned characteristics. Then explain how Colleen might have developed or learned that characteristic or trait.

Colleen's learned characteristic or trait is

_____

_____

How did Colleen develop or learn this characteristic or trait?

_____

_____

_____

_____

_____

_____

Step 1

Read the scenario "Producers and Consumers."

Producers and Consumers

All animals depend on plants for their food. Plants are called producers because they can make their own food. Some animals, like rabbits and mice, eat only plants for food. Other animals, like wolves and people, eat animals that eat the plants. If an organism cannot make its own food and eats plants or other animals for food, it is called a consumer.

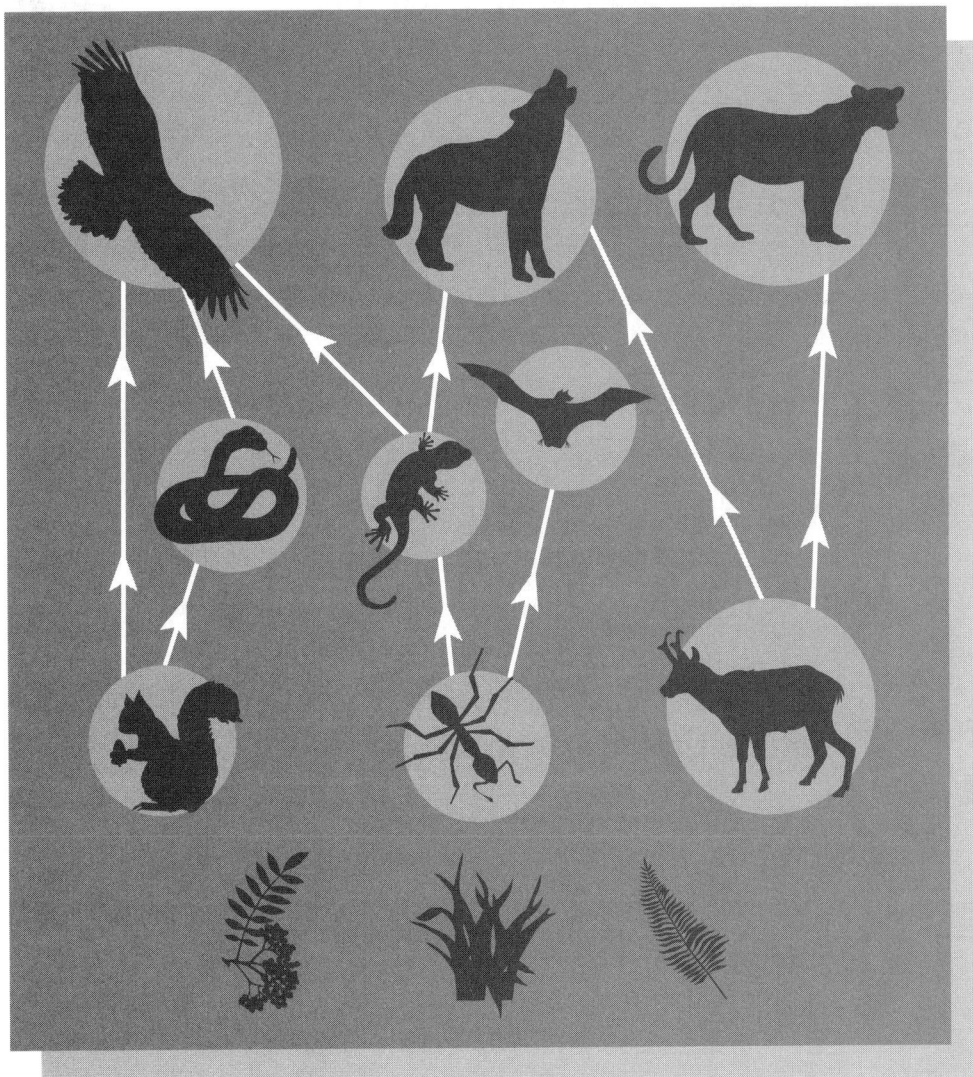

Step 2

Complete the Checklist "Clues for Success."

The checklist will help you to read and think like a scientist.

Clues for Success

☐ **C**arefully read the information.

☐ **L**ook at any illustrations or diagrams.
     They may provide you with additional information to answer the question.

☐ **U**nderstand the way you are asked to answer the question.
     ☐ Graph
     ☐ Chart
     ☐ Diagram
     ☐ Complete sentences
     ☐ Phrase
     ☐ Filled circle

☐ **E**xamine the information given.
     ☐ Reread the questions.
     ☐ Underline key words or phrases.
     ☐ Think about what the questions are asking.

☐ **S**ee if your answers match the questions.

Step 3

Use the information from
"Producers and Consumers"
to complete the graphic organizer.

Read the list of plants and animals on the table below. Then decide if
the organism is a producer (can make its own food) or consumer (must
consume a plant or organism that ate a plant). Completely fill the circle
in the space that shows your answer choice.

	producer	consumer
grass	◯	◯
chicken	◯	◯
flowers	◯	◯
people	◯	◯
snake	◯	◯
trees	◯	◯
grasshopper	◯	◯
rabbit	◯	◯
mice	◯	◯
berry bush	◯	◯
hawk	◯	◯
wheat	◯	◯
fox	◯	◯
cow	◯	◯

Activity 5

Step 4

Answer the following question for "Producers and Consumers" using information from your graphic organizer.

1. A food chain shows one pathway for food. Use the organisms from the table on page 89 to create a three-organism food chain that ends with people. Be sure to name each organism and use arrows to show the flow of energy.

Step 1

Read the scenario
"Kelp Forest Environment."

Kelp Forest Environment

An organism's patterns of behavior are connected to the organism's environment. This includes the kinds and numbers of other organisms present and the availability of food and resources. When the environment changes, some plants and animals survive and reproduce, while others die or move on to new locations. An example of this interaction is the kelp forest.

Sea otters live in patches of kelp. Kelp is a large, brown-green seaweed (plant) that grows in shallow ocean waters. Kelp provides food and shelter for many animals. Kelp helps the aquatic animals hide from their predators. If the kelp would disappear, these creatures would not have protection. The sea otters wrap themselves in kelp to keep from floating away. Sea otters help the kelp by eating the animals that feed on the kelp. One of the creatures that the otters eat is the sea urchin. When the sea urchins nibble on the kelp, the kelp can get loose from the ocean floor and die.

Step 2

Complete the Checklist "Clues for Success."

The checklist will help you to read and think like a scientist.

Clues for Success

☐ **C**arefully read the information.

☐ **L**ook at any illustrations or diagrams.
　　They may provide you with additional information to answer the question.

☐ **U**nderstand the way you are asked to answer the question.
　　☐ Graph
　　☐ Chart
　　☐ Diagram
　　☐ Complete sentences
　　☐ Phrase
　　☐ Filled circle

☐ **E**xamine the information given.
　　☐ Reread the questions.
　　☐ Underline key words or phrases.
　　☐ Think about what the questions are asking.

☐ **S**ee if your answers match the questions.

Step

3

Use the information from
"Kelp Forest Environment"
to complete the graphic organizer.

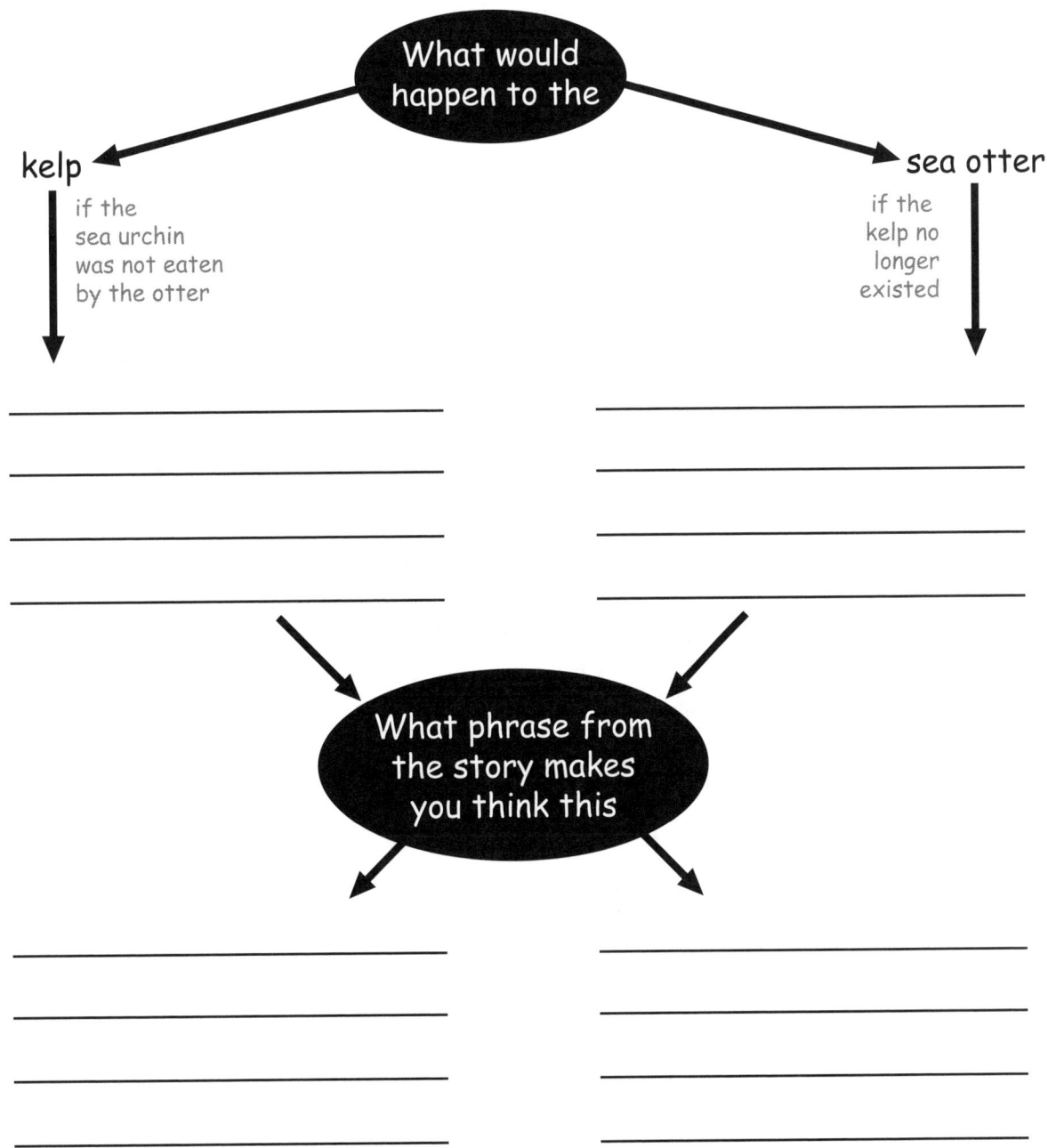

What would
happen to the

kelp   sea otter

if the
sea urchin
was not eaten
by the otter

if the
kelp no
longer
existed

_____   _____

_____   _____

_____   _____

_____   _____

What phrase from
the story makes
you think this

_____   _____

_____   _____

_____   _____

_____   _____

Step 4

Answer the following questions for "Kelp Forest Environment" using information from your graphic organizer.

1. Does this paragraph describe a land or water environment?

_____

2. What word or phrase from the paragraph makes you think this?

_____

_____

3. What is the producer in this environment?

_____

4. What word or phrase from the paragraph makes you think this?

_____

_____

© Englefield & Associates, Inc.

Step 1

Read the scenario "Skateboard Debate."

Skateboard Debate

A town council is having a meeting to discuss the idea of building a new skateboarding park near the school. The students are excited about the project and think it would be beneficial for the town. Some other members of the community think this would not be a good decision. The students made a list of the top three reasons they thought it would be a good idea. The parents and community members also made a list of their top three reasons they like or do not like the idea.

Students' List

___ There will be another fun activity

___ More skateboarders will come to the area

___ Can walk to the skate park

Parents' List

___ Children can walk to the park.

___ New and old friends can meet at the park.

___ Children will get more exercise.

___ Some of the trees around the school will be cut down.

___ Skateboarders will bring more noise to the neighborhood.

___ Skateboarders will bring more trash to the neighborhood.

Community's List

Step 2

Complete the Checklist "Clues for Success."

The checklist will help you to read and think like a scientist.

Clues for Success

☐ **C** arefully read the information.

☐ **L** ook at any illustrations or diagrams.
They may provide you with additional information to answer the question.

☐ **U** nderstand the way you are asked to answer the question.
  ☐ Graph
  ☐ Chart
  ☐ Diagram
  ☐ Complete sentences
  ☐ Phrase
  ☐ Filled circle

☐ **E** xamine the information given.
  ☐ Reread the questions.
  ☐ Underline key words or phrases.
  ☐ Think about what the questions are asking.

☐ **S** ee if your answers match the questions.

Step 3	Use the information from "Skateboard Debate" to complete the graphic organizer.

Read each of the lists on page 95. Place a **+** next to the ideas on the lists on page 95 you think are good things and a **−** next to the ideas on the list you think are problems. Write two good things (benefits) and two problems (detriments) with the skateboarding park idea on the lines below.

Good Things (benefits)

1. _____

_____

2. _____

_____

Problems (detriments)

1. _____

_____

2. _____

_____

Step 4

Answer the following question for "Skateboard Debate" using information from your graphic organizer.

1. Think about the animals that live in the woods near the school. Do you think the skateboarding park will help or hurt the animals? Use a completely filled circle to show your answer choice. Then write two reasons that support your choice.

I think the animals will be

○ helped   ○ hurt

because

reason one   reason two

_____   _____
_____   _____
_____   _____
_____   _____
_____   _____

Chapter 4

The activities in this section of the book will focus on Earth Science.

These activities will focus on:

- the properties of materials found on Earth such as rocks, soil, water, and gases,

- objects that are in the sky, and

- changes that impact Earth and the sky such as weather and changes to Earth's surface.

Use the "Clues for Success" Checklists as you complete each activity in this section as a tool to help you do your best work.

Step 1

Read the scenario
"Earth Materials."

Earth Materials

Earth provides materials for people to use as they meet their needs for life.

Some of the materials are provided in the atmosphere. These are gases like the oxygen people breathe and the carbon dioxide plants use when they produce food.

The water that is necessary for life is provided by the hydrosphere. Water can be found in rivers, lakes, oceans, streams, and under the ground.

The lithosphere provides soil for growing food; rocks, stones, and materials for building; and fuels like oil, coal, and petroleum to provide energy resources.

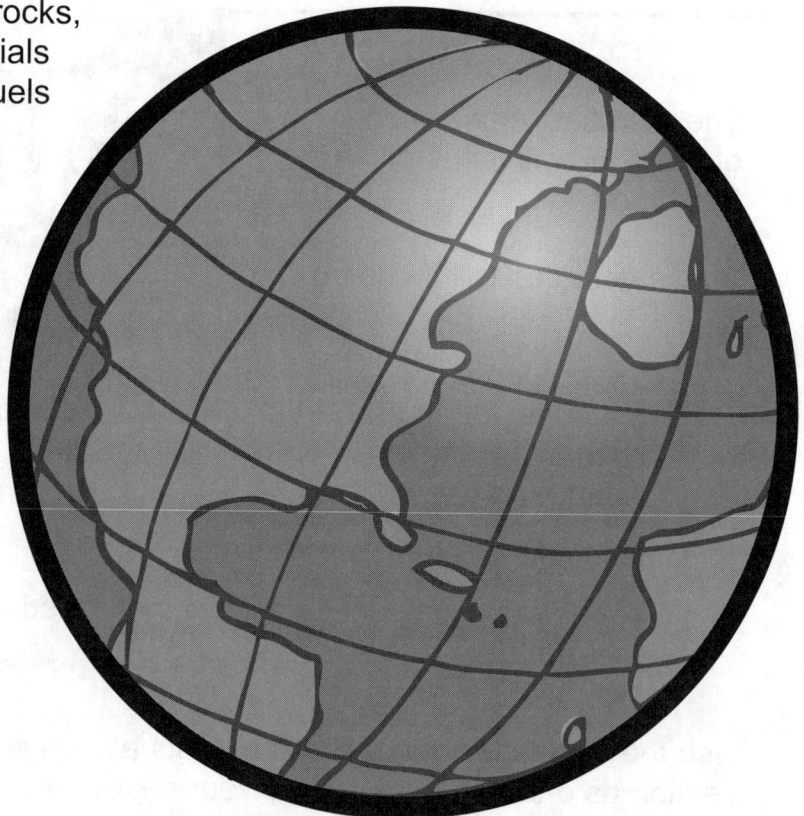

Step 2

Complete the Checklist "Clues for Success."

The checklist will help you to read and think like a scientist.

Clues for Success

☐ **C**arefully read the information.

☐ **L**ook at any illustrations or diagrams.
They may provide you with additional information to answer the question.

☐ **U**nderstand the way you are asked to answer the question.
　☐ Graph
　☐ Chart
　☐ Diagram
　☐ Complete sentences
　☐ Phrase
　☐ Filled circle

☐ **E**xamine the information given.
　☐ Reread the questions.
　☐ Underline key words or phrases.
　☐ Think about what the questions are asking.

☐ **S**ee if your answers match the questions.

Step 3

Use the information from "Earth Materials" to complete the graphic organizer.

Some of the materials that people use are

☐ rocks   ☐ stones   ☐ coal   ☐ petroleum   ☐ water

☐ oxygen   ☐ carbon dioxide   ☐ soil

Complete the flow chart to show how each of the materials listed above are used by people. Check off each material when you use the word for the chart.

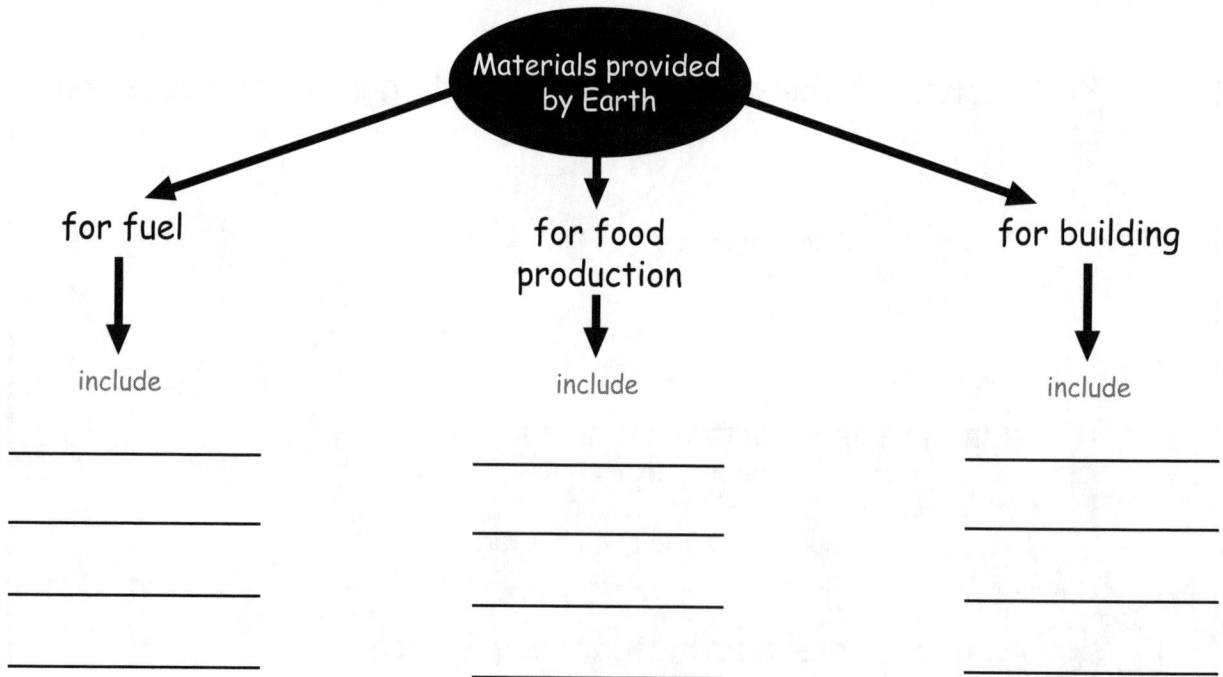

Materials provided by Earth

for fuel   for food production   for building

include   include   include

_____   _____   _____

_____   _____   _____

_____   _____   _____

Step 4

Answer the following questions for "Earth Materials" using information from your graphic organizer.

1. Materials from Earth are considered precious by people. These materials include metals such as gold and silver. Other precious materials include gems such as diamonds and emeralds.

   How do people use these precious materials?

   _____

   _____

   _____

   _____

   How does the way people use precious materials differ from the way people use needed materials from Earth?

   _____

   _____

   _____

   _____

   _____

Step

1

Read the scenario
"Coal Types."

Coal Types

The fourth grade class is learning about the four kinds of coal that are mined for use by people.

- **Peat** is a brown crumbly material that contains a lot of water. The water must be removed before peat can be used as a fuel.

- **Lignite** is a brown coal. Lignite is not a good fuel because it does not give off a lot of heat.

- **Bituminous** coal is a good fuel that burns at a high temperature. Bituminous coal contains sulfur. The sulfur in this coal harms the environment because it pollutes the air.

- **Anthracite** is a hard, black coal. It contains more carbon than any other coal. It burns hot and gives off a small amount of smoke and ash as it burns.

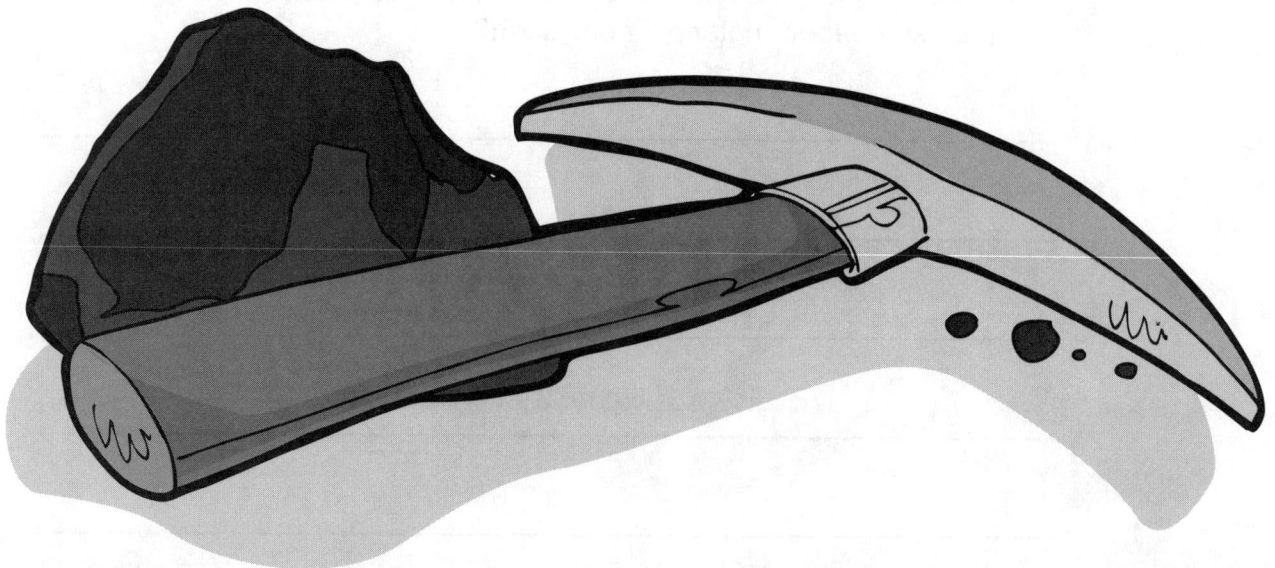

Step 2

Complete the Checklist "Clues for Success."

The checklist will help you to read and think like a scientist.

Clues for Success

☐ **C**arefully read the information.

☐ **L**ook at any illustrations or diagrams.
　　They may provide you with additional information to answer the question.

☐ **U**nderstand the way you are asked to answer the question.
　　☐ Graph
　　☐ Chart
　　☐ Diagram
　　☐ Complete sentences
　　☐ Phrase
　　☐ Filled circle

☐ **E**xamine the information given.
　　☐ Reread the questions.
　　☐ Underline key words or phrases.
　　☐ Think about what the questions are asking.

☐ **S**ee if your answers match the questions.

Step 3

Use the information from "Coal Types" to complete the graphic organizer.

The students created a chart to help them learn the characteristics about each type of coal. Complete the chart with at least two qualities that will help you describe the type of coal listed above the column.

Peat	Bituminous

Lignite	Anthracite

Step 4

Answer the following questions for "Coal Types" using information from your graphic organizer.

1. Coal is a fuel that was formed from plants that lived millions of years ago. When the plants died they sank to the bottom of swamps that covered the earth at the time. As the years passed the plants were buried deeper into the earth. Pressure changed the plant remains by forcing water out. Heat from the earth and more pressure formed a hard coal that is used as fuel. Coal is mined from the earth using equipment so it can be used by people.

   Read each of the sentences about coal formation. Number the statements from 1 to 8 to describe the sequence of how plants were formed into the fuel, coal.

   _____ Plants grew on the earth.

   _____ People use equipment to mine coal from the earth.

   _____ The mined coal is used as a fuel by people.

   _____ Dead plants sank into the swamps that covered the earth.

   _____ The growing plants died.

   _____ The water in the plants was forced out by the pressure surrounding them.

   _____ Heat and pressure on the plant changed the plant into coal.

   _____ As years passed, the plants were buried deep in the earth's swamps.

Step 1

Read the scenario "Terrific Terrariums."

Terrific Terrariums

The students were making terrariums at the environmental conference. Each student got a clear one-liter container with a lid. Here is the list of supplies on the lab table.

Clear plastic containers with lids

Potting soil—provides nutrients, allows water to drain

Clay—does not allow water to drain

Vermiculite—mix with soil, helps hold (retain) water in the soil

Pebbles, sand, rocks—help with drainage

Sand—to mix in for special succulent plants like cacti

Plant food—mix with water before adding to the container

Water—needed for plants to grow

Josh volunteered to show the class how he would make a terrarium. He selected the following for his demonstration:
    vermiculite
    potting soil
    pebbles
    ivy plant
    plant food mixed in water

Step 2

Complete the Checklist "Clues for Success."
The checklist will help you to read and think like a scientist.

Clues for Success

☐ **C** arefully read the information.

☐ **L** ook at any illustrations or diagrams.
They may provide you with additional information to answer the question.

☐ **U** nderstand the way you are asked to answer the question.
- ☐ Graph
- ☐ Chart
- ☐ Diagram
- ☐ Complete sentences
- ☐ Phrase
- ☐ Filled circle

☐ **E** xamine the information given.
- ☐ Reread the questions.
- ☐ Underline key words or phrases.
- ☐ Think about what the questions are asking.

☐ **S** ee if your answers match the questions.

Step 3

Use the information from "Terrific Terrariums" to complete the graphic organizer.

Josh's note cards fell and they were out of order when he gathered them. Look at the diagram on page 108 to help Josh put the note cards in the correct order, from 1 (first step) to 8 (last step) so he can do his best.

_____ Put pebbles in the bottom of the container for drainage.

_____ Spoon the soil mixture into the container.

_____ Mix vermiculite and soil together.

_____ Mix the water and plant food together.

_____ Water the plant before putting the lid on the terrarium.

_____ Tightly, fasten the lid.

_____ Take the completed terrarium home and give to my grandmother.

_____ Plant the ivy.

 © Englefield & Associates, Inc.

Step 4

Answer the following questions for "Terrific Terrariums" using information from your graphic organizer.

1. Josh did not choose clay when he made his terrarium. Why do you think Josh did not use the clay in his terrarium?

_____

_____

_____

_____

2. Josh put the lid on the container after he completed the planting and watering. Do you think the plant will grow with the lid on the container? Why or why not?

_____

_____

_____

_____

Step

1

Read the scenario "Fossils Facts."

Fossil Facts

The class was learning about fossils. They learned that fossils provide evidence of plants and animals that lived long ago. They also learned that fossils give information about the area where they were found.

Two of the students in the class shared their experiences with fossils.

Jake told the class that his family visited the La Brea Tar Pits in California. The tar pits were filled with very sticky material. His family saw many fossils that were found in the area. The fossils included bones or prints of bones from deer, elephants, and saber-toothed tigers.

Mandy brought in a fossil she bought at the museum gift shop. It was a piece of glassy-like yellow material with a fly stuck inside of it. She said that the material was amber, a sticky sap material from a pine tree.

Step 2

Complete the Checklist "Clues for Success."

The checklist will help you to read and think like a scientist.

Clues for Success

☐ **C**arefully read the information.

☐ **L**ook at any illustrations or diagrams.
  They may provide you with additional information to answer the question.

☐ **U**nderstand the way you are asked to answer the question.
  ☐ Graph
  ☐ Chart
  ☐ Diagram
  ☐ Complete sentences
  ☐ Phrase
  ☐ Filled circle

☐ **E**xamine the information given.
  ☐ Reread the questions.
  ☐ Underline key words or phrases.
  ☐ Think about what the questions are asking.

☐ **S**ee if your answers match the questions.

Step 3

Use the information from "Fossils Facts" to complete the graphic organizer.

Use a Venn diagram to show how the fossil experiences that Jake and Mandy shared were alike and different.

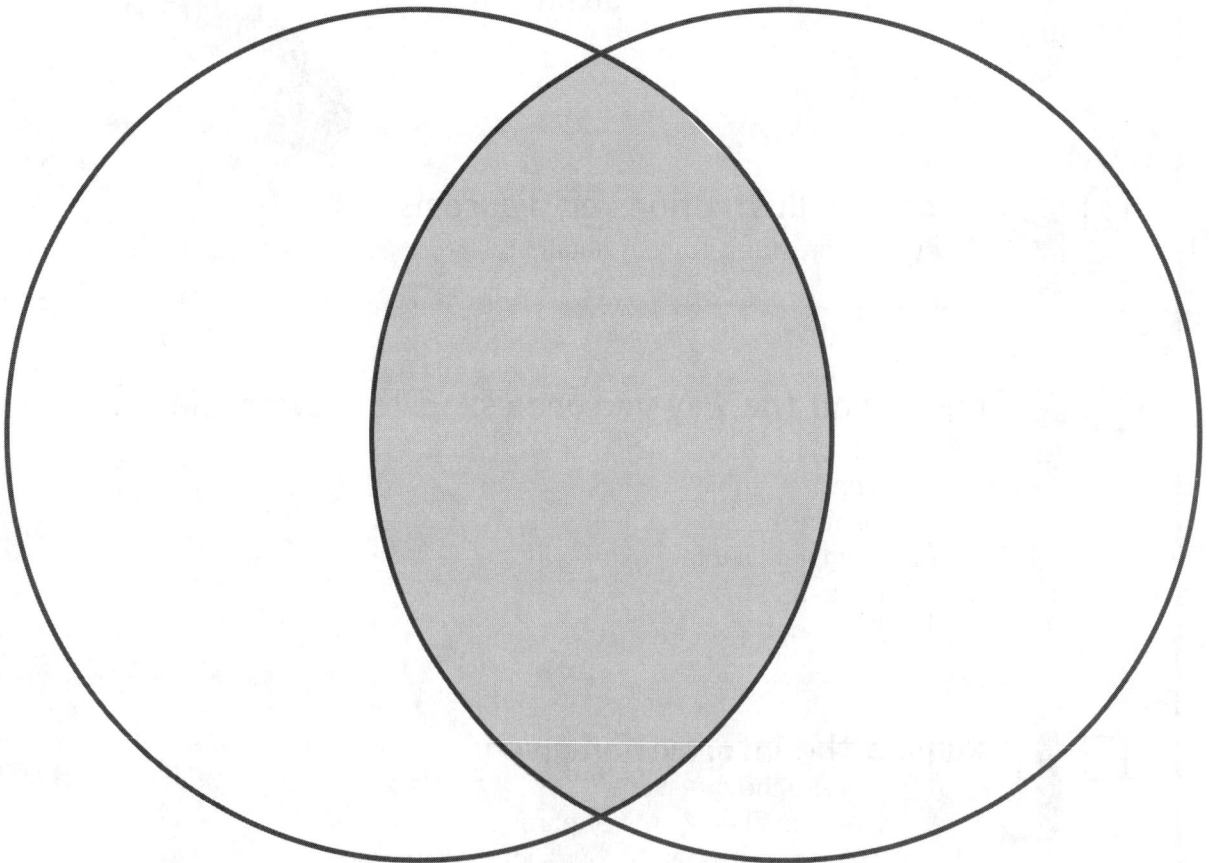

Step 4

Answer the following questions for "Fossils Facts" using information from your graphic organizer.

1. What information does Jake's fossil experience give about the La Brea Tar Pits?

_____

_____

_____

_____

_____

2. What information does Mandy's fossil give about where it was formed?

_____

_____

_____

_____

_____

Step 1

Read the scenario
"Sky High."

Sky High

The fourth grade students were studying about objects they could see in the sky. At first, they only thought of objects they could see in the day time. Below is their first word bank.

kite	sun	birds
airplane	clouds	

The teacher reminded them that they should include objects in the night sky. Here is their completed word wall.

kite	sun	birds
airplane	clouds	other planets
the moon	stars	satellite

Step 2

Complete the Checklist "Clues for Success."

The checklist will help you to read and think like a scientist.

Clues for Success

☐ **C**arefully read the information.

☐ **L**ook at any illustrations or diagrams.
　　They may provide you with additional information to answer the question.

☐ **U**nderstand the way you are asked to answer the question.
- ☐ Graph
- ☐ Chart
- ☐ Diagram
- ☐ Complete sentences
- ☐ Phrase
- ☐ Filled circle

☐ **E**xamine the information given.
- ☐ Reread the questions.
- ☐ Underline key words or phrases.
- ☐ Think about what the questions are asking.

☐ **S**ee if your answers match the questions.

Step 3

Use the information from "Sky High" to complete the graphic organizer.

Look at the word wall the students created listing the objects they could see in the sky. Complete the chart below using words from their word bank.

I provide Earth with warmth.	
I change my appearance in a 28-day cycle.	
I carry moisture and scatter rain, snow, hail, and sleet.	
I am an organism that can move in the sky with my own power.	
I fly using engines.	
I seem to twinkle in the night sky.	
I have a cloth tail and need wind to fly.	
I am an object in the solar system similar to Earth.	
I am a man-made object in orbit around Earth.	

Step 4

Create a dichotomous key for objects that move in the sky using the information from "Sky High" and your graphic organizer.

1. Write the words describing natural and man-made objects on the lines provided. Then for each category below them, write the word described in the shaded box. Rewrite the other words on the spaces provided.

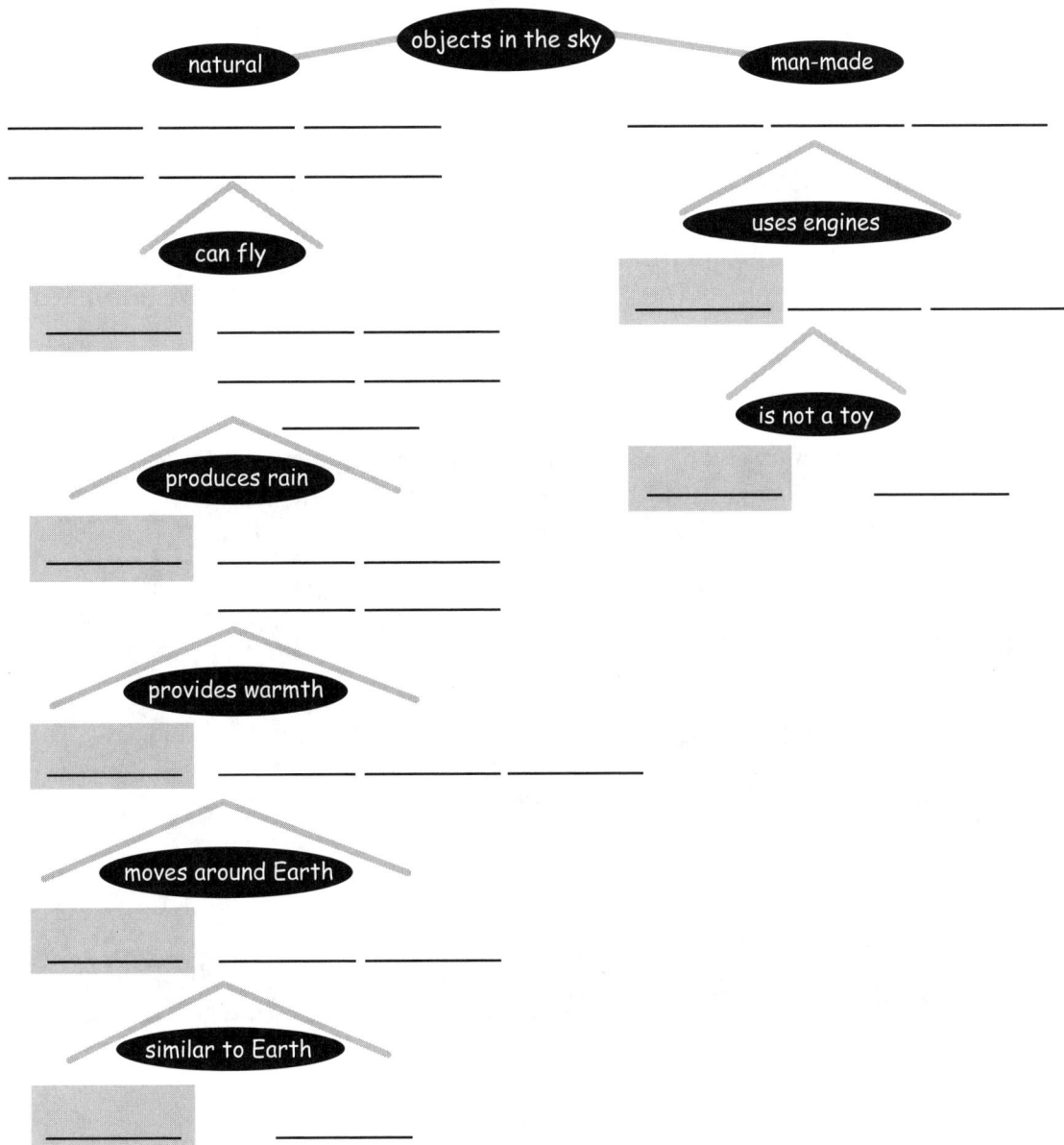

objects in the sky

natural   man-made

_____ _____ _____   _____ _____ _____

_____ _____ _____

can fly   uses engines

_____   _____ _____   _____   _____ _____

_____ _____

produces rain   is not a toy

_____   _____ _____   _____   _____

_____ _____

provides warmth

_____   _____ _____ _____

moves around Earth

_____   _____ _____ _____

similar to Earth

_____   _____

Step 1

Read the scenario
"The Sun: A Source of Heat and Light."

The Sun: A Source of Heat and Light

The sun is a large star that is the center of the solar system. The sun provides light and heat to Earth. The revolution of Earth around the sun and the rotation of Earth on its axis as it revolves around the sun changes the amount of light and the amount of heat Earth receives from the sun.

As Earth revolves around the sun, the sun's rays strike Earth's surface and atmosphere. Most of this energy is absorbed by Earth's surface.

The amount of light Earth receives from the sun as it rotates on its axis determines the amount of daylight and night experienced at a particular time.

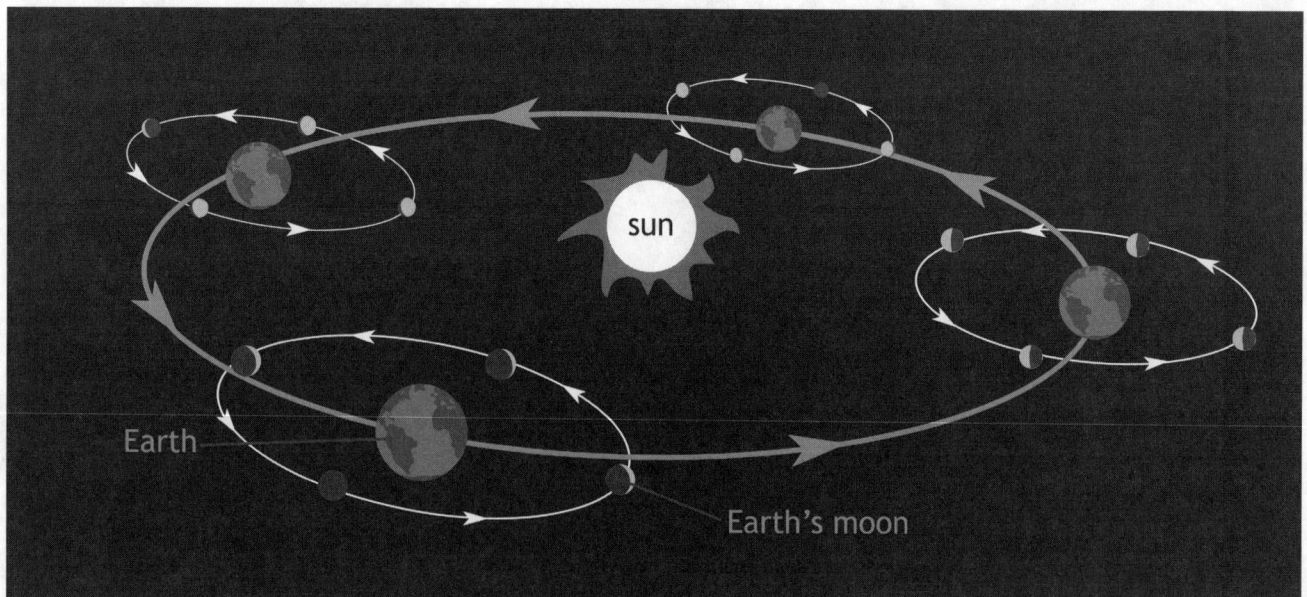

Step 2

Complete the Checklist "Clues for Success."

The checklist will help you to read and think like a scientist.

Clues for Success

☐ **C**arefully read the information.

☐ **L**ook at any illustrations or diagrams.
They may provide you with additional information to answer the question.

☐ **U**nderstand the way you are asked to answer the question.
  ☐ Graph
  ☐ Chart
  ☐ Diagram
  ☐ Complete sentences
  ☐ Phrase
  ☐ Filled circle

☐ **E**xamine the information given.
  ☐ Reread the questions.
  ☐ Underline key words or phrases.
  ☐ Think about what the questions are asking.

☐ **S**ee if your answers match the questions.

Step 3

Use the information from "The Sun: A Source of Heat and Light" to complete the graphic organizer.

Read each statement, then decide if you agree or disagree with it. Use a completely filled circle to show your choice. Then write the reason for your answer in the space provided.

The sun's **heat** makes the amount of daylight longer in the summer.

○ agree   ○ disagree

Reason:

_____

_____

The amount of sunlight makes summer days **warm** for swimming.

○ agree   ○ disagree

Reason:

_____

_____

Step 4

Answer the following question for "The Sun: A Source of Heat and Light" using information from your graphic organizer.

1. Read each statement. Decide if the statement is true or false. Write true or false on the lines provided. Then underline the words that helped you to decide if the statement was true or false.

_____ The sun's heat makes the morning light shine in the bedroom window each day.

_____ The amount of the sun's heat received makes the days seem longer in the winter.

_____ The amount of the sun's light will create the frigid temperatures that close school in the winter.

_____ The amount of light received from the sun helps create fun winter sports like sledding.

_____ The amount of sunlight makes December 20th the shortest day of the year.

Step 1

Read the scenario
"Earth Changes."

Earth Changes

The surface of Earth is always changing. Some of the changes are due to slow processes and others are due to rapid processes.

The slow processes take place over a long period of time. They are usually not noticed until the result is visible. Some of the slow processes include:

- a tree's roots breaking through a sidewalk,
- wind eroding (wearing away) a statue in the park, and
- large ruts forming in the soil where people ride dirt bikes.

The fast processes occur quickly over a short period of time. They are noticed as soon as they affect the area. Some of the fast processes that change Earth's surface include:

- an earthquake,
- a landslide,
- a volcano, and
- an avalanche.

               © Englefield & Associates, Inc.

Step 2

Complete the Checklist "Clues for Success."

The checklist will help you to read and think like a scientist.

Clues for Success

☐ **C**arefully read the information.

☐ **L**ook at any illustrations or diagrams.
  They may provide you with additional information to answer the question.

☐ **U**nderstand the way you are asked to answer the question.
  ☐ Graph
  ☐ Chart
  ☐ Diagram
  ☐ Complete sentences
  ☐ Phrase
  ☐ Filled circle

☐ **E**xamine the information given.
  ☐ Reread the questions.
  ☐ Underline key words or phrases.
  ☐ Think about what the questions are asking.

☐ **S**ee if your answers match the questions.

Step 3

Use the information from "Earth Changes" to complete the graphic organizer.

Select one kind of process that changes Earth's surface.

_____

Is this a slow change or a fast change? Use a completely filled circle to show your answer choice.

○ Slow change   ○ Fast change

Create a drawing or cartoon to illustrate your choice.

Step 4

Answer the following question for "Earth Changes" using information from your graphic organizer.

1. Describe how the action in your drawing or cartoon illustrates the slow or fast change on the surface of Earth. Be sure to use complete sentences.

_____

_____

_____

_____

_____

_____

_____

_____

_____

Step 1

Read the scenario "A Change in the Weather."

A Change in the Weather

Weather changes from day to day and season to season. These changes can be measured and described using a variety of tools. Three of these tools include:

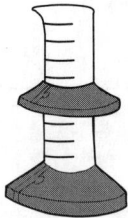

Rain gauge—A rain gauge collects the amount of rain that falls. It has marks and measurements to help you see how much rain fell.

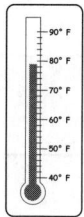

Thermometer—A thermometer measures the temperature. A thermometer can be digital so you can easily read the numbers. A thermometer may be a tube of colored liquid that rises as the temperature increases or goes down as the temperature decreases. You tell the temperature by reading the number at the height of the liquid.

Weather vane—A weather vane moves to point out the direction of the wind; North, South, East, and West. The speed of the wind can be shown by the movement of the pointer on the weather vane.

Step 2

Complete the Checklist "Clues for Success."

The checklist will help you to read and think like a scientist.

Clues for Success

☐ **C**arefully read the information.

☐ **L**ook at any illustrations or diagrams.
They may provide you with additional information to answer the question.

☐ **U**nderstand the way you are asked to answer the question.
 - ☐ Graph
 - ☐ Chart
 - ☐ Diagram
 - ☐ Complete sentences
 - ☐ Phrase
 - ☐ Filled circle

☐ **E**xamine the information given.
 - ☐ Reread the questions.
 - ☐ Underline key words or phrases.
 - ☐ Think about what the questions are asking.

☐ **S**ee if your answers match the questions.

Step 3

Use the information from
"A Change in the Weather"
to complete the graphic organizer.

A student prepares index cards to help him describe to the class how to use these weather tools. Complete each of the cards to help the student. In the space provided, explain how to use each of the tools.

Rain Gauge

What does it measure?

How do I measure with it?

         © Englefield & Associates, Inc.

Thermometer

What does it measure?

How do I measure with it?

Weather Vane

What does it show?

How do I measure with it?

Step 4

Answer the following riddles for "A Change in the Weather" using information from your graphic organizer.

You can use me to notice how fast the wind is blowing.

What am I?

_____

You use me to see the amount of water that fell on the garden during the night.

What am I?

_____

You can use me to see if you should wear a hat and gloves to go outside.

What am I?

_____

You will use me to see the direction of the wind.

What am I?

_____

© Englefield & Associates, Inc.

Step 1

Read the scenario "Seasonal Weather Game."

Seasonal Weather Game

The students were playing a game about the four seasons. They had to toss a bean bag onto a mat with the names of the seasons written on it.

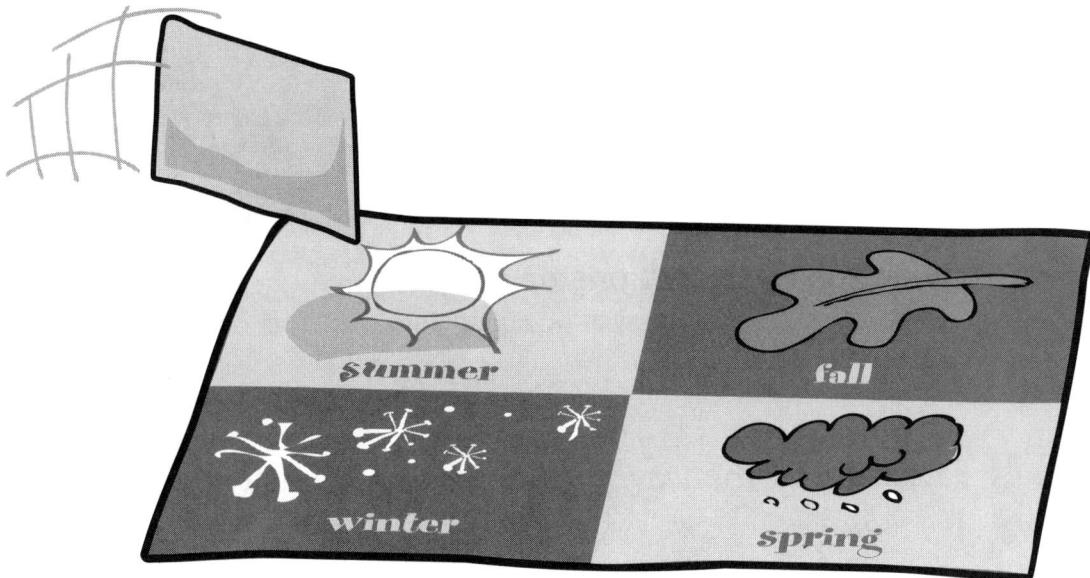

summer

fall

winter

spring

Then the students had to describe how the weather during this season would create changes in the conditions for people, plants, or playground equipment.

Step 2

Complete the Checklist "Clues for Success."

The checklist will help you to read and think like a scientist.

Clues for Success

☐ **C**arefully read the information.

☐ **L**ook at any illustrations or diagrams.
They may provide you with additional information to answer the question.

☐ **U**nderstand the way you are asked to answer the question.
- ☐ Graph
- ☐ Chart
- ☐ Diagram
- ☐ Complete sentences
- ☐ Phrase
- ☐ Filled circle

☐ **E**xamine the information given.
- ☐ Reread the questions.
- ☐ Underline key words or phrases.
- ☐ Think about what the questions are asking.

☐ **S**ee if your answers match the questions.

Step 3

Use the information from
"Seasonal Weather Game"
to complete the graphic organizer.

Hillary's bean bag landed on Spring. In the spaces provided, describe two seasonal weather changes (effects) people, plants, or playground equipment experience in the spring.

Spring Changes

people	1.
	2.
plants	1.
	2.
playground equipment	1.
	2.

Step 4

Answer the following question for "Seasonal Weather Game" using information from your graphic organizer.

1. What is your favorite season of the year?

Create a journal entry that has a symbol for this season and use complete sentences to describe two changes you experience in this season.

     © Englefield & Associates, Inc.

Chapter 5

The activities in this section of the book will focus on Science and Technology.

These activities will help you develop abilities to:

- communicate, design and propose a solution to a question,

- understand that many different kinds of people do Science and create technology to learn about the world and make life easier, and

- know the difference between natural and man-made objects.

Use the "Clues for Success" Checklists as you complete each activity in this section as a tool to help you do your best work.

Step 1

Read the scenario
"A Just Ducky Invention."

A Just Ducky Invention

Ariella was watching the ducks at the metro park with her mother when she noticed that the water was not sticking to the duck's body. In fact, the water looked like little beads on the feathers and the water would drop off when the duck moved. This gave Ariella an idea for her Science invention project, "Waterproof Gloves."

When she got home, she looked up "ducks" on the Internet and learned that oil is produced from a special gland to help make ducks' feathers waterproof. This is a special adaptation because ducks spend so much time in the water.

Ariella experimented with oil to help her invent waterproof gloves.

With her mother's permission, Ariella got the equipment she needed ready on the kitchen table.

Ariella carefully wrote out a plan.

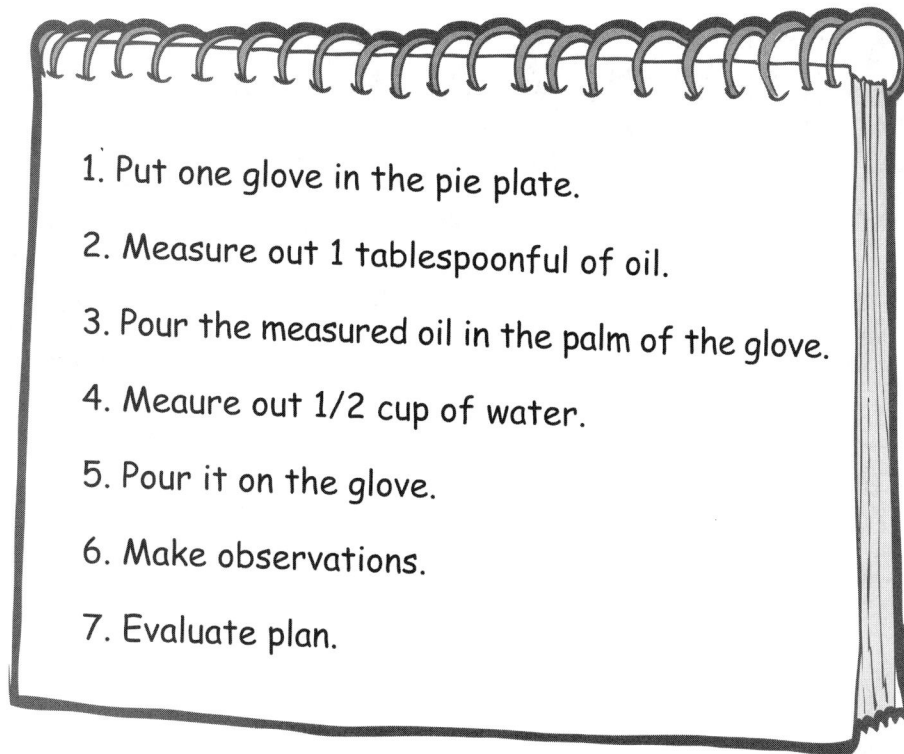

1. Put one glove in the pie plate.

2. Measure out 1 tablespoonful of oil.

3. Pour the measured oil in the palm of the glove.

4. Meaure out 1/2 cup of water.

5. Pour it on the glove.

6. Make observations.

7. Evaluate plan.

Ariella's Observation

When Ariella poured water on the glove the water rolled off the oil spot, but the rest of the glove got wet and soggy.

Activity 1

Step 2

Complete the Checklist "Clues for Success."

The checklist will help you to read and think like a scientist.

Clues for Success

☐ **C**arefully read the information.

☐ **L**ook at any illustrations or diagrams.
They may provide you with additional information to answer the question.

☐ **U**nderstand the way you are asked to answer the question.
- ☐ Graph
- ☐ Chart
- ☐ Diagram
- ☐ Complete sentences
- ☐ Phrase
- ☐ Filled circle

☐ **E**xamine the information given.
- ☐ Reread the questions.
- ☐ Underline key words or phrases.
- ☐ Think about what the questions are asking.

☐ **S**ee if your answers match the questions.

Step 3

Use the information from
"A Just Ducky Invention"
to complete the graphic organizer.

What is one problem with Ariella's waterproof gloves?

_____

_____

Suggest a solution Ariella might use to improve her invention of
waterproof gloves.

_____

_____

_____

_____

_____

_____

Step 4

Answer the following questions for "A Just Ducky Invention" using information from your graphic organizer.

1. Did Ariella's experiment show that the oil would make the glove waterproof?

   ○ yes   ○ no

   Why or why not?

   _____

   _____

   _____

2. Do you think Ariella's classmates will want to wear her invention, "Waterproof Gloves"?

   ○ yes   ○ no

   Why or why not?

   _____

   _____

   _____

Step 1

Read the scenario "Bunch of Bubbles."

Bunch of Bubbles

Simone's group was planning their Science fair project. They decided that they would design an experiment to determine which liquid detergent solution would make the best bubbles. Betsy volunteered to keep a notebook with all of the team ideas.

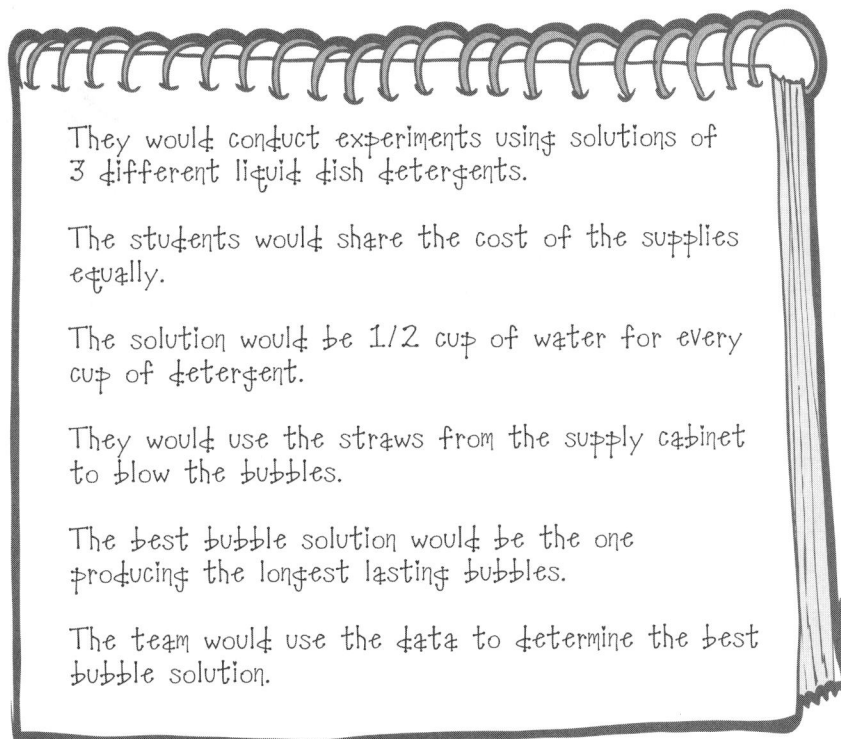

> They would conduct experiments using solutions of 3 different liquid dish detergents.
>
> The students would share the cost of the supplies equally.
>
> The solution would be 1/2 cup of water for every cup of detergent.
>
> They would use the straws from the supply cabinet to blow the bubbles.
>
> The best bubble solution would be the one producing the longest lasting bubbles.
>
> The team would use the data to determine the best bubble solution.

Harper was going shopping with her parents, so she offered to purchase the various detergents and other supplies.

Simone was elected to blow the bubbles because she was famous for her ability to blow bubble gum bubbles.

Jasmine was the data collector.

Step 2

Complete the Checklist "Clues for Success."

The checklist will help you to read and think like a scientist.

Clues for Success

☐ **C**arefully read the information.

☐ **L**ook at any illustrations or diagrams.
　　They may provide you with additional information to answer the question.

☐ **U**nderstand the way you are asked to answer the question.
　　☐ Graph
　　☐ Chart
　　☐ Diagram
　　☐ Complete sentences
　　☐ Phrase
　　☐ Filled circle

☐ **E**xamine the information given.
　　☐ Reread the questions.
　　☐ Underline key words or phrases.
　　☐ Think about what the questions are asking.

☐ **S**ee if your answers match the questions.

　　　　© Englefield & Associates, Inc.

Step 3

Use the information from
"Bunch of Bubbles" to complete
the graphic organizer.

Show each student's contribution to the team project.

Student	Contribution to the Project
Simone	
Betsy	
Harper	
Jasmine	

Step 4

Answer the following question for "Bunch of Bubbles" using information from your graphic organizer.

Use complete sentences to list three things this experimental design illustrates about a group project in Science.

1. _____

   _____

   _____

2. _____

   _____

   _____

3. _____

   _____

   _____

  © Englefield & Associates, Inc.

Step 1

Read the scenario "Science Bulletin Board."

Science Bulletin Board

Women and men of all ages and backgrounds engage in scientific and technological work.

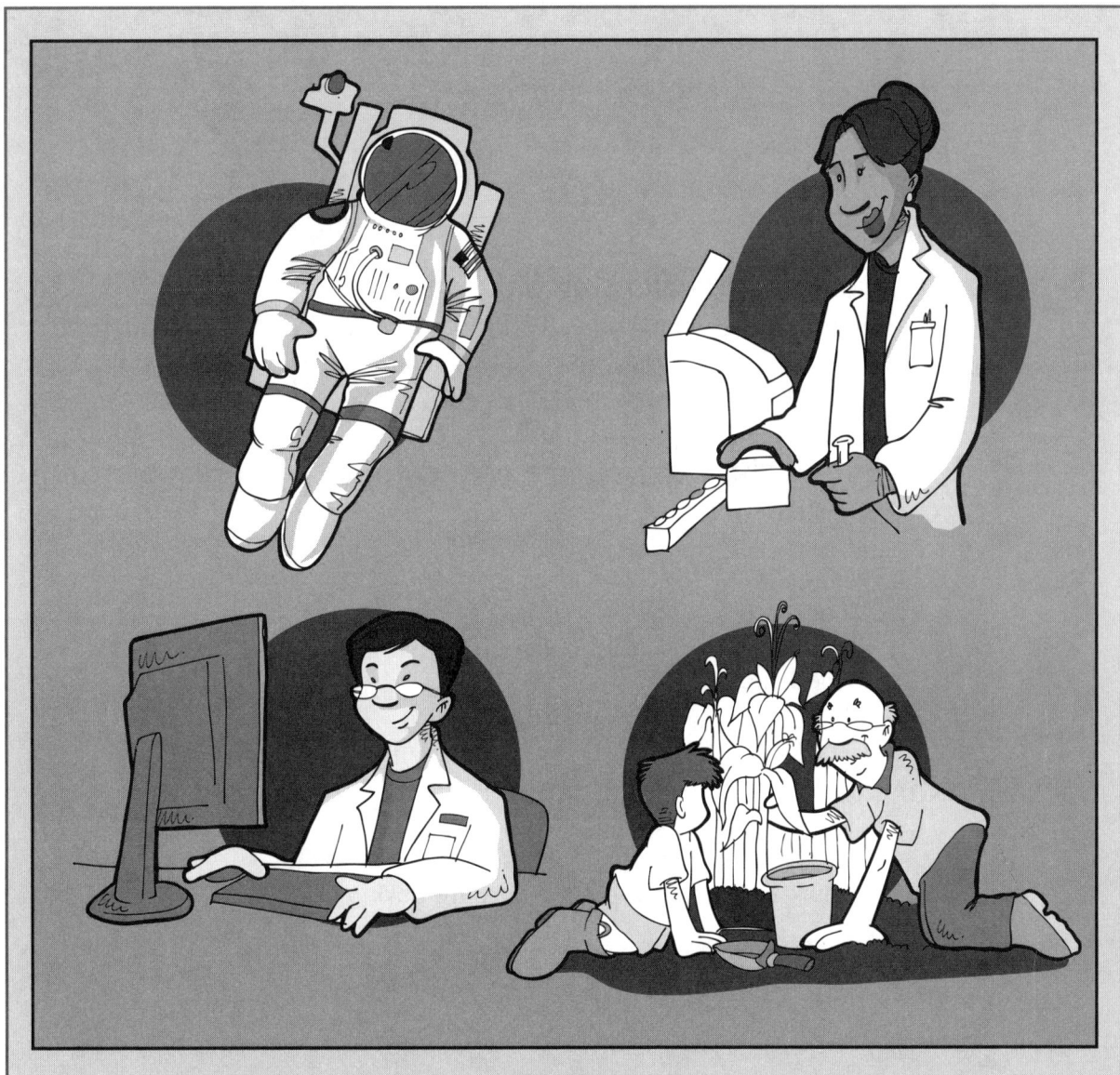

Step 2

Complete the Checklist "Clues for Success."
The checklist will help you to read and think like a scientist.

Clues for Success

☐ **C**arefully read the information.

☐ **L**ook at any illustrations or diagrams.
They may provide you with additional information to answer the question.

☐ **U**nderstand the way you are asked to answer the question.
☐ Graph
☐ Chart
☐ Diagram
☐ Complete sentences
☐ Phrase
☐ Filled circle

☐ **E**xamine the information given.
☐ Reread the questions.
☐ Underline key words or phrases.
☐ Think about what the questions are asking.

☐ **S**ee if your answers match the questions.

© Englefield & Associates, Inc.

Step **3**	Use the information from "Science Bulletin Board" to complete the graphic organizer.

Look at the pictures of scientists on the bulletin board. Then read each of the statements below. Use a completely filled circle to indicate if the statement is true or false about scientists.

	true	false
Men and women are scientists.	◯	◯
Scientific processes can only occur in a lab.	◯	◯
Science happens outdoors.	◯	◯
All scientists are adults.	◯	◯
Science can only occur indoors.	◯	◯
Science research can be done on a computer.	◯	◯
Science must be done with special tools.	◯	◯
Scientists are college graduates.	◯	◯

Activity 3

Answer the following questions for "Science Bulletin Board" using information from your graphic organizer.

1. Write a sentence about the different kinds of people who are involved in doing Science.

_____

_____

_____

_____

_____

2. Write a sentence about the work that is considered Science.

_____

_____

_____

_____

Step 1

Read the scenario "Natural or Man-Made?"

Natural or Man-Made?

People appreciate and use many types of objects in their lives. Some objects are designed in nature and others are designed by people.

Objects designed by nature are generally referred to as "natural." An example of natural objects would be a feather, shell, or a plant as they are found in nature.

Objects designed by people are referred to as "man-made." An example of a man-made object would be a silk plant that looks like a real plant. Or a man-made object like a shell necklace that uses a natural object in a way that is different from how it is used in nature. A man-made object can also be inspired by nature. For example, the way a burr (a natural object) sticks to items to spread seeds was the basis for the invention of Velcro®.

Step 2

Complete the Checklist "Clues for Success."

The checklist will help you to read and think like a scientist.

Clues for Success

- ☐ **C**arefully read the information.

- ☐ **L**ook at any illustrations or diagrams.
  They may provide you with additional information to answer the question.

- ☐ **U**nderstand the way you are asked to answer the question.
  - ☐ Graph
  - ☐ Chart
  - ☐ Diagram
  - ☐ Complete sentences
  - ☐ Phrase
  - ☐ Filled circle

- ☐ **E**xamine the information given.
  - ☐ Reread the questions.
  - ☐ Underline key words or phrases.
  - ☐ Think about what the questions are asking.

- ☐ **S**ee if your answers match the questions.

Step 3

Use the information from
"Natural or Man-Made?"
to complete the graphic organizer.

Look at each set of objects. Use a completely filled circle to show the natural object. Use complete sentences to describe how the man-made objects were inspired by the natural object.

objects	how the man-made objects were inspired by the natural object

Step 4

Answer the following questions for "Natural or Man-Made?" using information from your graphic organizer.

Circle the natural objects. Then write a complete sentence to describe how the natural object is different from the man-made object.

1.

_____

_____

_____

2.

_____

_____

_____

 © Englefield & Associates, Inc.

Chapter 6

The activities in this section of the book will focus on Science in Personal and Social Perspective.

The activities in this section will help you focus on:

- personal health,
- characteristics and changes in populations,
- resources,
- changes in environments, and
- challenges in local environments.

Science in Personal and Social Perspective

Use the "Clues for Success" Checklists as you complete each activity in this section as a tool to help you do your best work.

Activity 1

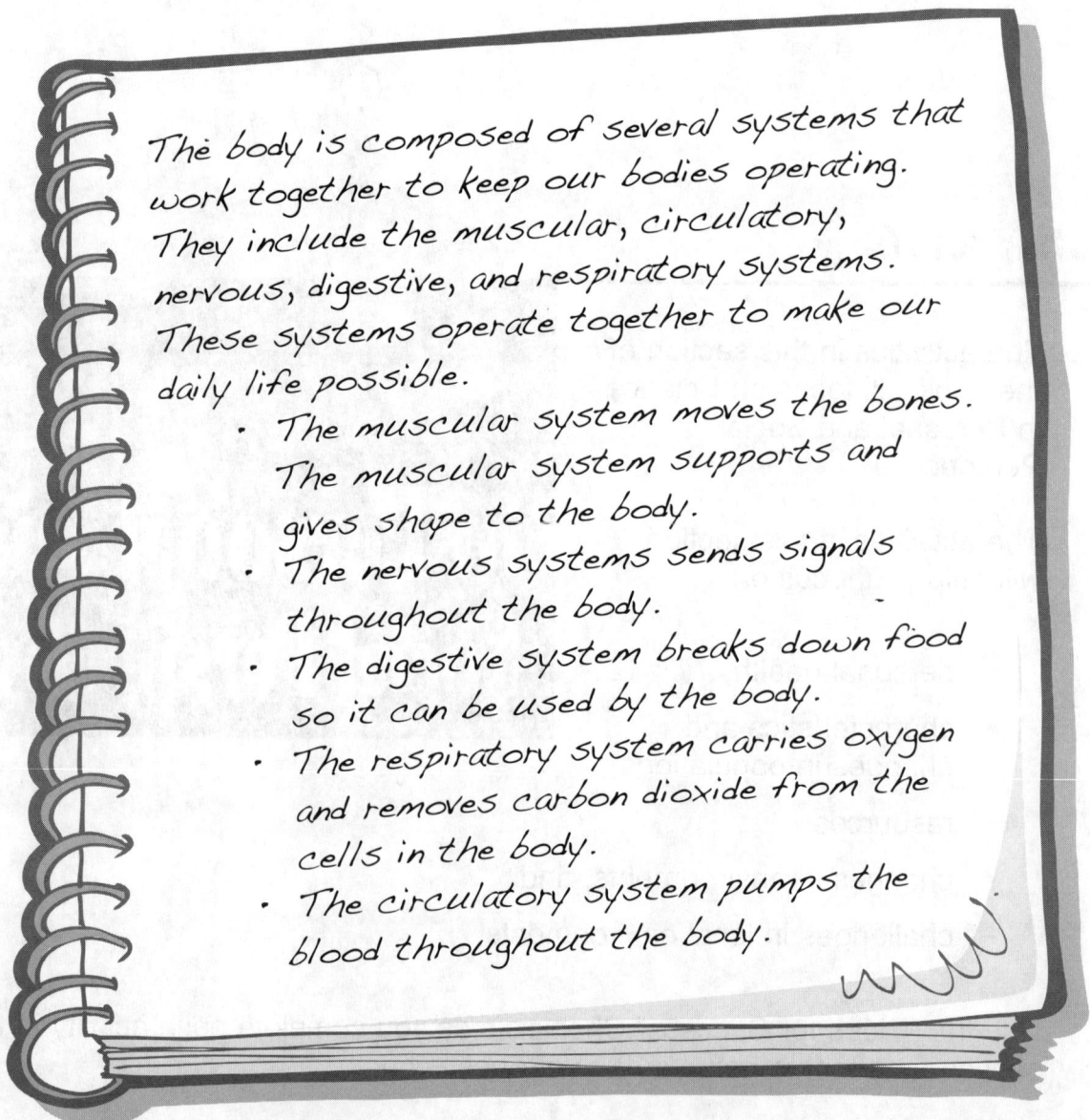

Step

1

Read the scenario
"Body Systems."

Body Systems

Susan took notes on the systems of the body when she was studying for her health quiz. Here is a page from her notebook.

The body is composed of several systems that work together to keep our bodies operating. They include the muscular, circulatory, nervous, digestive, and respiratory systems. These systems operate together to make our daily life possible.

- The muscular system moves the bones. The muscular system supports and gives shape to the body.
- The nervous systems sends signals throughout the body.
- The digestive system breaks down food so it can be used by the body.
- The respiratory system carries oxygen and removes carbon dioxide from the cells in the body.
- The circulatory system pumps the blood throughout the body.

Step 2

Complete the Checklist "Clues for Success."

The checklist will help you to read and think like a scientist.

Clues for Success

☐ **C**arefully read the information.

☐ **L**ook at any illustrations or diagrams.
    They may provide you with additional information to answer the question.

☐ **U**nderstand the way you are asked to answer the question.
- ☐ Graph
- ☐ Chart
- ☐ Diagram
- ☐ Complete sentences
- ☐ Phrase
- ☐ Filled circle

☐ **E**xamine the information given.
- ☐ Reread the questions.
- ☐ Underline key words or phrases.
- ☐ Think about what the questions are asking.

☐ **S**ee if your answers match the questions.

Activity 1

Step 3

Use the information from "Body Systems" to complete the graphic organizer.

Use the page from the notebook on page 156 to complete the table below. Look at the organ listed and then identify the body system in which it works.

Organ	Body System
Heart	
Lungs	
Stomach	
Brain	
Arm Muscles	
Nose	
Blood Vessels	

 © Englefield & Associates, Inc.

Step 4

Answer the following question for "Body Systems" using information from your graphic organizer.

1. Describe how the sense organ shown would affect the body system given.

	digestive system

	muscular system

	nervous system

Step 1

Read the scenario
"Healthy Skating."

Healthy Skating

For her research project in health class, Jordan wanted to learn more about fitness activities. Jordan enjoys ice skating, so she wants her report to be about the sport of ice skating. The teacher said the students must use four or more resources as a requirement for the assignment.

To start the assignment, Jordan made an outline.

My favorite sport is ice skating

I. Different types of ice skating I could try
    A. Speed skating
    B. Figure skating

II. My body and ice skating
    A. Muscles used in ice skating
    B. How to take care of sore muscles
    C. How to care for sprained ankles

III. Conclusion

Step 2

Complete the Checklist "Clues for Success."
The checklist will help you to read and think like a scientist.

Clues for Success

☐ **C**arefully read the information.

☐ **L**ook at any illustrations or diagrams.
    They may provide you with additional information to answer the question.

☐ **U**nderstand the way you are asked to answer the question.
    ☐ Graph
    ☐ Chart
    ☐ Diagram
    ☐ Complete sentences
    ☐ Phrase
    ☐ Filled circle

☐ **E**xamine the information given.
    ☐ Reread the questions.
    ☐ Underline key words or phrases.
    ☐ Think about what the questions are asking.

☐ **S**ee if your answers match the questions.

Step 3

Use the information from "Healthy Skating" to complete the graphic organizer.

Next to the topic on Jordan's outline write the name of a resource from the list that might help her find the information she needs for each part of her report. Be sure to use four of the resources listed. You may use a resource more than once.

computer   gym teacher   doctor
school nurse   encyclopedia   expert at skating rink

My favorite sport is ice skating

I. Different types of
    ice skating I could try _____
        A. Speed skating _____
        B. Figure skating _____

II. My body and ice skating
        A. Muscles used in
            ice skating _____
        B. How to take care
            of sore muscles _____
        C. How to care for
            sprained ankles _____

Step 4

Answer the following questions for "Healthy Skating" using information from your graphic organizer.

1. Read over the outline Jordan made for her report. Jordan does not have an idea for an ending to her report. Write an idea Jordan could include for the conclusion to her report, then name a resource Jordan might use to get information on that topic.

Conclusion Idea

_____

Resource

_____

2. Select one of these ideas or write one of your own on the space provided. Then give a resource that could help Jordan find information on the topic.

○   Famous Olympic skaters

○   Diet and calorie use for an ice skater

○   The best sport equipment to prevent injuries

Resource

_____

Activity 3

Step 1

Read the scenario "Flooded Field Trip."

Flooded Field Trip

The students were disappointed to learn that the field trip to the metro park was cancelled because the rain storm flooded the area. The park ranger came to the school to tell the students why the area floods during a rain storm. In her presentation, the ranger showed aerial photographs (pictures taken from the sky) of the park area as it looked 20 years ago and the way it looks today. She explained that 20 years ago flooding was not a problem for the park but now it is a problem.

20 years ago

Today

Step 2

Complete the Checklist "Clues for Success."

The checklist will help you to read and think like a scientist.

Clues for Success

☐ **C**arefully read the information.

☐ **L**ook at any illustrations or diagrams.
　　They may provide you with additional information to answer the question.

☐ **U**nderstand the way you are asked to answer the question.
　　☐ Graph
　　☐ Chart
　　☐ Diagram
　　☐ Complete sentences
　　☐ Phrase
　　☐ Filled circle

☐ **E**xamine the information given.
　　☐ Reread the questions.
　　☐ Underline key words or phrases.
　　☐ Think about what the questions are asking.

☐ **S**ee if your answers match the questions.

Step 3

Use the information from "Flooded Field Trip" to complete the graphic organizer.

Use the T-Chart to compare the area on the photographs. List three ways the area stayed the same. List three ways the area changed.

Area on Photographs	
ways area stayed the same	ways area changed
1.	1.
2.	2.
3.	3.

Step 4 — Answer the following questions for "Flooded Field Trip" using information from your graphic organizer.

Use the information you listed about how the metro park area is different to answer the following questions.

1. Select one change that you listed that helped (benefited) the people that live in the area. Then use a complete sentence to explain how the change helped the people.

   Change: _____

   How this helped the people

   _____

   _____

2. Select one change that you listed that might have contributed to the flooding (harmed) of the park. Then use a complete sentence to explain how the change contributed to the flooding in the park.

   Change: _____

   How this contributed to the flooding in the park

   _____

   _____

Chapter 7

The activities in this section of the book will focus on the History and Nature of Science.

The activities in this section will help you learn that:

- Science is created and conducted by people.

- Science has been practiced by people for a long time.

- Men and women choose careers in Science and make contributions in the areas of Science and Technology.

- Scientists are always adding to what people are learning about the world.

- Science will never be finished.

Use the "Clues for Success" Checklists as you complete each activity in this section as a tool to help you do your best work.

History & Nature of Science

Step 1

Read the scenario
"Researching the Light Bulb."

Researching the Light Bulb

Martin prepared a report on the invention of the light bulb. He used the dates and the countries of scientific discoveries that lead to the present day light bulb. His research showed that Science and Technology have been practiced by people for a long time.

1800—England

Humphry Davy was experimenting with electricity and connected wires to a battery and a piece of carbon. This connection made the carbon glow producing light called an electric arc. This was the first electric light.

1877—America

Charles Francis Brush manufactured some carbon arcs to light a public square in Cleveland, Ohio, USA. These light arcs were used on streets, in large office buildings, and in stores.

1878—England

Sir Joseph Wilson Swan produced a practical, long-lasting electric light using a carbon paper filament (thin wire). This filament worked well but burned up quickly. He demonstrated his new electric lamps in Newcastle, England.

      © Englefield & Associates, Inc.

1879—America

Thomas Alva Edison experimented with thousands of different filaments to find just the right materials to glow and be long-lasting. Edison discovered that a carbon filament in an oxygen-free bulb glowed for 40 hours but did not burn up.

1881—America

Lewis Howard Latimer improved the bulb by inventing a carbon filament. In 1882, he developed and patented a method of manufacturing his carbon filaments.

1903—America

Willis R. Whitney invented a treatment for the filament so that it wouldn't darken the inside of the bulb with soot as it glowed.

1910—America

William David Coolidge invented a tungsten filament which lasted longer than the older filaments.

Step 2

Complete the Checklist "Clues for Success."

The checklist will help you to read and think like a scientist.

Clues for Success

☐ **C**arefully read the information.

☐ **L**ook at any illustrations or diagrams.
They may provide you with additional information to answer the question.

☐ **U**nderstand the way you are asked to answer the question.
   ☐ Graph
   ☐ Chart
   ☐ Diagram
   ☐ Complete sentences
   ☐ Phrase
   ☐ Filled circle

☐ **E**xamine the information given.
   ☐ Reread the questions.
   ☐ Underline key words or phrases.
   ☐ Think about what the questions are asking.

☐ **S**ee if your answers match the questions.

Step 3

Use the information from "Researching the Light Bulb" to complete the graphic organizer.

Complete the chart to identify the scientist and country of origin for the changes to the light bulb described.

change to light bulb	scientist	country
carbon paper filament		
filament treated so it would not darken glass		
invented tungsten filament		

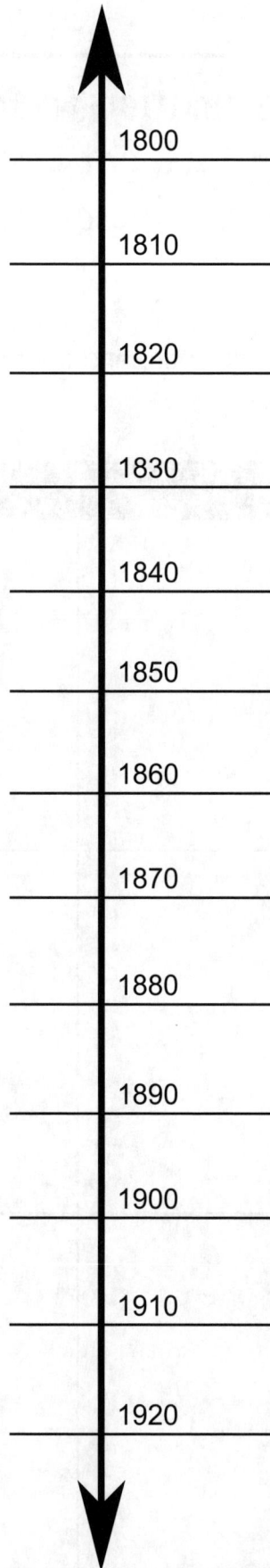

1800

1810

1820

1830

1840

1850

1860

1870

1880

1890

1900

1910

1920

Step 4

Answer the following questions for "Researching the Light Bulb" using information from your graphic organizer.

1. Use the information on Martin's cards to help make a timeline of the following changes in the development of the light bulb.

   Cut and glue each statement to its corresponding place on the timeline on page 174.

   > Thomas Alva Edison was experimenting with a carbon filament and a oxygen free bulb.

   > A patent was obtained for manufacturing carbon filaments.

   > Carbon arcs were used to light streets in Cleveland, Ohio.

   > An electric arc was produced with a battery and a piece of carbon.

   > Tungsten filaments were invented.

   > Filaments do not create soot inside the bulb.

   > The first electric light was created.

After Martin's presentation, he asked the class to answer these questions. Answer each question with a complete sentence.

1. How is the work of Charles Francis Brush connected to the work of Humphry Davy?

_____

_____

_____

_____

2. How did the contribution of William David Coolidge change the light bulb?

_____

_____

_____

_____

Step 1

Read the information about light bulb development.

The history of the development of the light bulb is an example of how scientific discoveries and inventions come about. The light bulb used in the world today is a modern version of an invention that reflects the work of many scientists, from different countries, over a long period of time.

History and Nature of Science

Parts of a Light Bulb include:

- The light is surrounded by a **thin glass bulb**.

- The bulb is filled with **non-reactive (inert) gases**.

- **Tungsten filament** is the coiled metal wire that glows brightly when electricity flows through it.

- **Support wires** hold up the filament.

- **Connecting wires** carry electricity from the bulb's electrical contact to the filament.

- **Electrical contacts** are located in the metallic base of the bulb. The two contacts connect to the ends of the electrical circuit.

- The **fuses** of the bulb (located in the stem of the bulb) are insulated by glass.

- The **cap** of the light bulb is located at the base of the bulb. It is the part that connects the bulb to the lamp.

Step 2

Complete the Checklist "Clues for Success."

The checklist will help you to read and think like a scientist.

Clues for Success

☐ **C**arefully read the information.

☐ **L**ook at any illustrations or diagrams.
    They may provide you with additional information to answer the question.

☐ **U**nderstand the way you are asked to answer the question.
    ☐ Graph
    ☐ Chart
    ☐ Diagram
    ☐ Complete sentences
    ☐ Phrase
    ☐ Filled circle

☐ **E**xamine the information given.
    ☐ Reread the questions.
    ☐ Underline key words or phrases.
    ☐ Think about what the questions are asking.

☐ **S**ee if your answers match the questions.

Step 3

Use the information about light bulb development to complete the graphic organizer.

Here is a model of the light bulb used today. Read about the parts of the light bulb on page 178. Use the definitions of the words in **bold letters** to help you label the parts of the light bulb.

© Englefield & Associates, Inc.

Step 4

Answer the following question using information about light bulb development and your graphic organizer.

Match each part of the light bulb in Column A with its use (function) in Column B.

Column A

_____ the glass bulb

_____ the connecting wires

_____ the cap

_____ the electric contacts

_____ the tungsten filament

Column B

A. connects the bulb to the lamp

B. is filled with inert (non-reactive) gases

C. glows brightly giving off light

D. located in the metallic base of the bulb

E. carries electricity to the filament

Step 1

Read the scenario "Science for Everyone."

Science for Everyone

David told the teacher that he thought only men were good scientists and inventors. The next day his teacher hung a group of posters in the classroom showcasing women who contributed knowledge and inventions in different areas. Here are the posters.

Marie Curie

Polish-born and French citizen physicist and chemist

Discovered the chemical elements of polonium and radium. With her husband Pierre coined the word radioactivity to describe the element radium.

Marie Curie won the Nobel Peace Prize twice. She is the only woman to win it in two different sciences.

© Englefield & Associates, Inc.

Ruth Graves Wakefield

American innkeeper of the Toll House Inn in Whitman, Massachusetts

Discovered broken chocolate bars would not melt in cookies when they baked in the oven.

Her invention was called the "Toll House Cookie" using the bits of chocolate now called chocolate chips.

Stephanie Louise Kwolek

American chemist

Discovered Kevlar

Kevlar is a material that is five times stronger than the same weight of steel.
Uses: Safety vests that could not be penetrated, helmets, trampolines, tennis rackets.

Beatrix Potter

English author and illustrator, botanist, and conservationist

Illustrated the images she saw with a microscope to show the interactions of organisms she found in nature (field and forest). This was the only way to record these relationships at the time.

Step 2

Complete the Checklist "Clues for Success."

The checklist will help you to read and think like a scientist.

Clues for Success

☐ **C**arefully read the information.

☐ **L**ook at any illustrations or diagrams.
　　They may provide you with additional information to answer the question.

☐ **U**nderstand the way you are asked to answer the question.
　　☐ Graph
　　☐ Chart
　　☐ Diagram
　　☐ Complete sentences
　　☐ Phrase
　　☐ Filled circle

☐ **E**xamine the information given.
　　☐ Reread the questions.
　　☐ Underline key words or phrases.
　　☐ Think about what the questions are asking.

☐ **S**ee if your answers match the questions.

　　© Englefield & Associates, Inc.

Step 3

Use the information from
"Science for Everyone"
to complete the graphic organizer.

Place the information from the posters into an outline form.

I. Marie Curie
   A. Country _____
   B. Contribution _____
   _____

II. Ruth Graves Wakefield
   A. Country _____
   B. Contribution _____
   _____

III. Stephanie Louise Kwolek
   A. Country _____
   B. Contribution _____
   _____

IV. Beatrix Potter
   A. Country _____
   B. Contribution _____
   _____

Activity 3

Step 4

Answer the following questions for "Science for Everyone" using information from your graphic organizer.

Carefully read the information on the posters. Use a complete sentence to answer the following questions using details from the posters.

1. Are all women scientists and inventors from America?

_____

_____

2. Are all women scientists and inventors chemists?

_____

_____

3. Do all inventions and scientific discoveries take place in a lab?

_____

_____

_____

Step 1

Read the scenario "Changes in Science."

Changes in Science

Science is the knowledge about the world around us and the way we get that knowledge. Tools change and this makes changes in our information. These tools include scientific information and scientific instruments. The information about the planets is a good example of how new information changes what scientists believe about the solar system.

Galileo was an Italian scientist who lived from 1564–1642. Galileo was the first scientist to observe the universe using a telescope. The development of the telescope helped make observations about objects that move in the solar system. Galileo's observations helped change an idea scientists believed at the time. Scientists believed that Earth was the center of the universe. Galileo's information showed that the sun is the center of the universe.

In 1905, Percival Lowell used mathematics to predict that something in the universe was affecting the orbits of Neptune and Uranus. In 1930, using the equipment in an observatory that Lowell built, Clyde Tombaugh took a photograph of a small object that he thought caused the change in the movement of Uranus and Neptune. This spot was accepted by the scientific community as the planet Pluto.

A change in information about the solar system occurred in 2006. At a meeting of the International Astronomical Union, scientists reviewed the information about Pluto's orbit and voted that Pluto was not a planet. They decided Pluto should be placed in a new category, Dwarf Planets.

These changes in knowledge about the solar system illustrate that although men and women have learned much about objects, events, and phenomena in nature, scientists are always gaining more information to understand. Science will never be finished.

Activity 4

Step 2

Complete the Checklist "Clues for Success."
The checklist will help you to read and think like a scientist.

Clues for Success

☐ **C**arefully read the information.

☐ **L**ook at any illustrations or diagrams.
    They may provide you with additional information to answer the question.

☐ **U**nderstand the way you are asked to answer the question.
    ☐ Graph
    ☐ Chart
    ☐ Diagram
    ☐ Complete sentences
    ☐ Phrase
    ☐ Filled circle

☐ **E**xamine the information given.
    ☐ Reread the questions.
    ☐ Underline key words or phrases.
    ☐ Think about what the questions are asking.

☐ **S**ee if your answers match the questions.

Step 3

Use the information from
"Changes in Science"
to complete the graphic organizer.

Complete the table with information from "Changes in Science."

	Scientists	Tool or Information Used	Change that Happened
1564–1642			
1930			
2006			

Step 4

Answer the following question for "Changes in Science" using information from your graphic organizer.

Read the statement.

Scientific knowledge about our world never changes.

1. Do you **agree** or **disagree** with the statement?

   ◯   ◯

   Give details from the scenario "Changes in Science" and your graphic organizer to explain your opinion.

   _____

   _____

   _____

   _____

 © Englefield & Associates, Inc.

Activity 5

Read the scenario
"Help Wanted: Science."

Help Wanted: Science

When the students studied Science careers, they created a word wall to help them remember the careers. They wrote their list on the chalkboard.

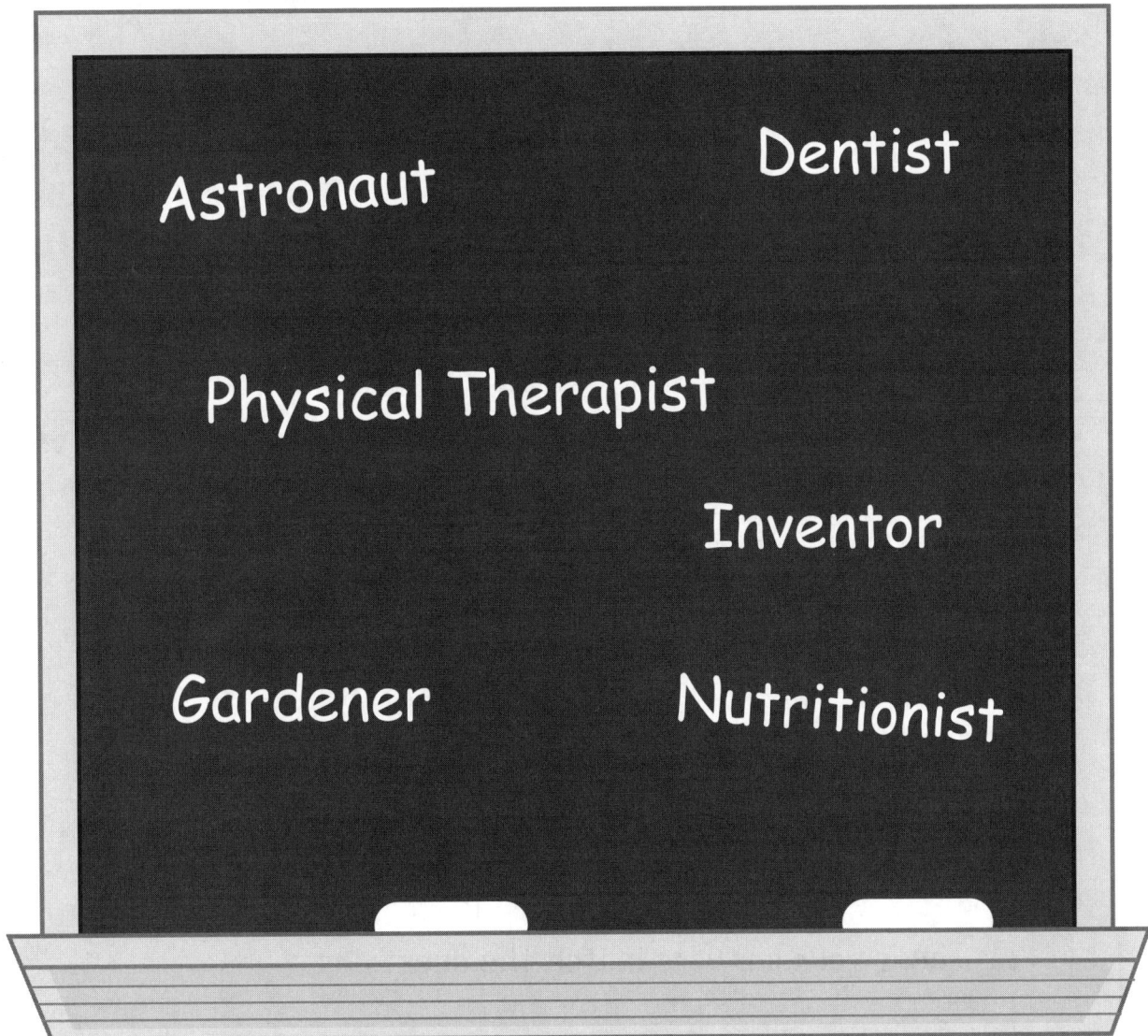

Astronaut   Dentist

Physical Therapist

                             Inventor

Gardener   Nutritionist

Step 2

Complete the Checklist "Clues for Success."

The checklist will help you to read and think like a scientist.

Clues for Success

☐ **C**arefully read the information.

☐ **L**ook at any illustrations or diagrams.
    They may provide you with additional information to answer the question.

☐ **U**nderstand the way you are asked to answer the question.
- ☐ Graph
- ☐ Chart
- ☐ Diagram
- ☐ Complete sentences
- ☐ Phrase
- ☐ Filled circle

☐ **E**xamine the information given.
- ☐ Reread the questions.
- ☐ Underline key words or phrases.
- ☐ Think about what the questions are asking.

☐ **S**ee if your answers match the questions.

 © Englefield & Associates, Inc.

Step 3

Use the information from "Help Wanted: Science" to complete the graphic organizer.

Read the types of different scientists listed on the word wall. Use your past experiences to write words or phrases to remind you about the kind of work each scientist does.

Physical Therapist

_____

Inventor

_____

Dentist

_____

Gardener

_____

Astronaut

_____

Nutritionist

_____

Step 4

Answer the following questions for "Help Wanted: Science" using information from your graphic organizer.

Read each of the help wanted ads below. Then select a scientist from the word wall to answer each help wanted ad.

HELP WANTED!
Adventurer needed to explore the planet, Mars.

1. Who will answer the ad?

_____

HELP WANTED!
Students want advice about taking care of their teeth.

2. Who will answer the ad?

_____

      © Englefield & Associates, Inc.

HELP WANTED!

The flowers have arrived!
Need help planting near the school entrance.

3. Who will answer the ad?

_____

HELP WANTED!

Ideas needed for new designs to make
students' book bags easier to carry.

4. Who will answer the ad?

_____

HELP WANTED!

Healthy meal ideas needed for
students at the Hometown School.

5. Who will answer the ad?

_____

HELP WANTED!

Sports team wants to learn how to strengthen muscles at the beginning of the season.

6. Who will answer the ad?

_____

© Englefield & Associates, Inc.